**Mankind cannot reproduce
pollen, honey or any honey
product . . . synthetically.**

FOREWORD

This book is a luminous landmark of historical significance of the remarkable qualities of bee pollen. It will enlighten many Americans about the health values of bee pollen. It is an excellent summary of the contemporary uses and values of bee pollen.

It is of inestimable value to those who have or will have the chronic degenerative diseases of diabetes or hypoglycemia. Many authorities believe that is has a value in the treatment of arthritis. It has been said that pollen helps to reduce the tendency for fatal fats to be deposited in the life line arteries and tends to reduce the incidence of artery disease. It also tends to minimize the ugly fat bulges when one over-indulges in overly processed carbohydrates.

It also stresses the value of the improved nutrient composition in natural foods of which bee pollen is one of the finest. It is encouraging that more and more people are becoming aware of what I said during an appearance on the TODAY Show — that most food in America will support life but it won't sustain health. Bee pollen is a very health giving food.

As a practicing proponent of total metabolic biocellular preventive health care it has been my privilege to observe and research and pursue many hours of diligent probing into the pursuit of optimum health. Since it is apparent that our health and our lives depend upon the quality of our cell function because cells make up our tissue, organs and glands, it seems then that we should advance a new theory of the chronic "breakdown diseases" — our major killers — to keep us from being sidetracked by the germ theory of disease.

Since we are realizing that if we have sick cells we have a sick body, we need to look to the environmental factors inside and outside our body we can help promote a better state of health and prevent the chronic breakdown diseases.

Since cells must have cell-food to be healthy, to form the basis for our health, it behooves us to look to the nutrient factors that are in our food or what I should term applied medical nutrition.

To help understand the application of total biocellular metabolic health care I have coined a phrase, the cell theory of disease, to better understand, treat and prevent the modern day breakdown diseases. As we study the cells needed nutrients it is apparent that we have a-nutrient-osis which signifies a lack of many nutrients in our food today replacing the old idea that a deficiency state as described as being a lack of one individual nutrient or a-vitamin-osis. This has been used to describe the frank deficiencies of a single vitamin.

And we have to coin another word to point out that we are adding more bodily damage by the effects of "anti-nutrition." This adds up all of the environmental factors that are dangerous and damaging to the amounts of proper nutrients left in our food and in our environment today. Anti-nutrition lessens the value of the nutrients left in our food as well as adding to the effects of the illness producing "chemical pollution."

Since sick cells will produce products of poor metabolism, this leads to a cell state that I like to call "Cell Pollution." Since all these factors lead to poor metabolic status in all cells, I prefer to call this state hypocelluosis, which of course is an indication of hypometabolic state with metabolic imbalances, which includes both anabolism - the building up and repairing and restoring part of metabolism - and the catabolic, which of course is the utilization part of our metabolism and constitutes the wear and tear of our daily metabolic existence.

Now we are beginning to have a basis for improved metabolism and a modern total metabolic health care approach to lessen the effects of our modern day affluent disease of civilization. This is why the health giving effects of bee pollen is so important in our work of good cell metabolism.

Joe Parkhill is a man of rare attributes. His tremendous intellectual curiosity has propelled him into the leadership of the apiary field. No one can deny that his outstanding sagacity about beekeeping and honey's outstanding healthful qualities has earned him his enviable especial title Honeyologist. Joe Parkhill has spent many years and traveled far and wide to gather all the valuable factual materials so ably presented in this book on the values of bee pollen.

Joe Parkhill has a remarkable capacity to present in layman language the scientific facts that he has gained throughout the world in other books on honey and this book on the marvelous food value of bee pollen.

Joe Parkhill is an outspoken exponent of the values of preventive medicine. It has been my privilege to give Mr. Parkhill an appreciation plaque from the International Preventive Medicine Foundation, which I had the honor of founding, for his support of the preventive medicine movement. The emergence of tranquilizers and anti-depressants which cover up and hide symptoms and do not treat basic metabolic processes have added more health problems. Antibiotics, the wonder drugs, have been a medical triumph but have failed to consider the basic problem of decreased health resistance and increased host acceptability when it occurs in most people today. It has been said that bee pollen has a great value in the prevention of some of the allergic and infective problems which are very common in America today.

The "search for truth is in one way hard, and in another easy." For it is evident that no one can master it fully nor miss it wholly, but each adds a little to our knowledge of nature, and from all these facts assembled, there arises a certain grandeur . . . Aristotle.

Dr. Richard O. Brennan
D.O., M.D., Ph.G, F.A.C.G.P.,
F.I.C.A.N., F.A.P.M., D.P.H.

APITHERAPY

For much too long we have let critics that know nothing about the subject belittle our experiences in the field of apitherapy. Some beekeepers are often "non-believers". In both cases it is because of a **lack of knowledge** and **experience.**

What wonderful "pharmacists" honeybees are. They work with Mother Nature, her plants and flowers, nourished by soil, water, air and sunshine, making the **greatest laboratory on earth,** to create all the infinite forms of life that exists today.

It is so encouraging to see such an increase of interest in the health value of bees, beekeeping, honeybee venom, pollen, propolis and all bee products.

Apitherapy, should once again replace other unnecessary therapeutics, that are **myths** and not **facts.**

Pollen and Honey, man's earliest and healthiest foods, have been valued throughout successive civilizations. Today, their popularity is increasing as never before.

Here is a lively and authoritative book covering the subjects of pollen, honey, propolis, royal jelly and beeswax.

POLLEN

How bee pollen controls weight. When bee pollen's inverted sugars are taken into your digestive system, there is a speedy combustion. This causes a speedy increase in the rate of calorie burning and consequent weight loss. It is the bee pollen which causes this internal reaction. It is nature's way to improve metabolism and to control, as well as take off, weight.

Bee pollen is the number one body builder and will not contribute fat to the body. Bee pollen is the number one food for weight reduction.

Honeybee pollen is an essential nutrient to put the body chemistry into balance and to rid itself of digestive problems.

NOTES

POLLEN

This book is meant to be in simple language that we may understand the wonders of the Honeybee and its health-giving products: Pollen — Honey — Propolis — Beeswax — Pollination — Royal Jelly.

For those who say, "We have no scientific facts!" . . . Honey, Pollen and **all** the honeybee products have been used since the beginning of time — ever since the world was created. All through Biblical times, the Roman Empire and in every country in the entire world up until this very day — Pollen and Honey have been known to be the most perfect food God has put on the earth.

It has **all** the vitamins, minerals, amino acids, everything the body must have to function. All these are in minute form . . . the way the body needs them. Honey even has 17% water which no other food has.

Pollen and Honey are the world's best preventive medicine. Pollen is not a drug but a food with no man-made chemical to upset your body in any way as long as it is honeybee pollen.

Stress occurs when you drain your body's physical resources by not eating or sleeping enough. It can cause muscle tension, change your breathing and heart rate, and make your body more susceptible to disease. Stress shows up as worry, guilt, anxiety and a desire to be alone.

Relaxing exercise, twice a day, helps prevent heart attacks by reducing blood pressure. Take 20 minutes of exercise by lying on your back on the bed - with legs spread a foot apart and your arms at your sides with your palms facing up - close your eyes and breathe slowly and deeply from your diaphram. Pause a second after inhaling and a second after exhaling. Then for 15 minutes relax the muscles of your body: First, those in the feet, then legs, stomach, lower back, chest, upper back, neck, face and head. In 3 to 4 weeks your blood pressure wil begin to drop. You are the one who can lower your blood pressure. If you are, or are not, taking medication . . . what do you have to lose?

Since the beginning of time, man has sought remedies and foods for enhancing his powers, when all the time God gave them to us in our food and drink. All we need to do is use our common sense. Take the "Pollen . . . the Honey" that he has given us and eat and enjoy them.

The popularity of Pollen is increasing as never before. So if you are looking for the world's oldest and original natural "picker-upper", all around preventive medicine and food . . . you will enjoy this book.

People have handed this important information and "facts" - not theory - down through the ages from Biblical times through the Roman Empire, through the Greek Regime to this present time.

These are facts handed down by physicians, Bible scholars, lawyers, people of all nations and all walks of life - including "Hippocrates" the father of modern medicine. His principles of medical science were laid down 400 years before the birth of Christ and they formed the basis for medical theory.

We now have ample proof of the health-giving properties of pollen.

The history of pollen leading up to its acceptance as a health food is an interesting study. Sweden leads the world in discoveries about pollen. From the time flowers appeared, there has been pollen. What is new about it is the fact that only during the last decade have scientists and doctors been able to accept and approve its true value.

The honeybee and primitive races of long ago have known and proven the value by using pollen as a food and a preventive medicine. Pollen again comes into the news because a Swedish beekeeper was clever enough to invent a pollen trap to get the bees to unwittingly harvest this food for man without hurting the bee or without leaving the bees short of pollen for themselves.

Pollen is a pure food and preventive medicine, nonhabit-forming, comparatively inexpensive, easy to obtain and easy to take. Make sure you get pure pollen — nothing added or nothing taken away.

Pollen has shown itself to be a complete nourishment in every sense of the word. It would be impossible to find a food from an animal or vegetable containing such vital nutritional elements. Not only does it build up strength and energy in tired bodies, but it also acts as a tonic. In most cases in ten days to two weeks people who have lost the zest in living have restored normal healthy appetites and a new outlook on life. Pollen will stimulate nearly all functions of the body, including the gastric system. Containing a natural antibiotic, it also controls bacteria in the intestines. The benefits experienced by persons taking pollen not only consisted of restored body health but also a more optimistic outlook on life and with more vigor, vitality and increased resistance to infection. Pollen acts as a tonic, rapidly restoring normal weight and energy to a convalescent.

The protein content of pollen compares to the fresh weight basis with that of seed and grain foodstuffs, as navy beans, peanuts and soybeans. (Titus, 1939) Vivins and Palmer (1944) found that protein, fat, phosphorous and iron content of mixed bee-collected Minnesota pollens resembles closely that of dried kidney and navy beans and dried peas, which help to show that pollen is indeed a wholesome food. (The Hive and the Honey Bee - Dadent & Sons, Inc.)

Deep research has proven pollen to be a food from natural organic sources. Pollen contains the richest source yet revealed of vitamins, minerals, proteins, amino acids, hormones, enzymes and fats. Like with honey, pollen is surrounded by certain mystery in that it contains other substances which defy identification. At the University of Arizona it was discovered that antibiotics were present in pollen.

Shapes and color indicate the species of plant from which it comes. Certain bees in the hive gather nectar to convert into honey, whereas others forage for pollen. Bees perform a remarkable, miraculous feat in the collection of pollen, for it is not swallowed into the honey sac like nectar. The pollen is worked into a mass and carried by their rear legs, which are expertly designed by nature. There are as many different tastes and colors of pollen pellets as there are flowers and plants.

Pollen is a highly concentrated substance which never contains more than 18 percent water, up to 35 percent protein, 15-25 percent amino acids and up to 40 percent carbohydrates or glucides, fruit sugar (cellulose), grape sugar (glucose). Total fats and oils are only about 5 percent on the average. Enzymes in pollen are amylase, catalase, dehydrogenase, diapase, diastase, cozymase, cytochrome pectasc, phosphatase, sucrase, lactic acids.

Pollen compares with yeast as a complete food when it is revealed that a mixture of fresh pollen can contain up to 1,000 micrograms of cozymase per gram; the alcoholic fermentation of yeast and pollen is identical.

All growing plants from the smallest to the largest shed millions of grains of pollen. The amount of pollen in the world is so enormous that the grains could be compared with the number of stars and the number of grains of sand in the world.

Pollen grains under certain conditions are undestructible and could prove that pollen grains are so intricately designed that under anaerobic conditions they are everlasting and totally immune from decay.

Without pollen there would be no plants, trees or flowers of any kind, and without these, man or beast could not survive.

NOTES

BEE POLLEN AND THE BODY

Many modern analysts, chemists, doctors and people the world over are finally learning that the ancient beliefs of preventive medicine and health problems are, and have always been, "facts" — not "theory".

Some say there are certain parts of the world where people live extraordinarily long lives. People in the United States or any country live long, healthy lives if they eat God-given food — such as bee pollen, honey, propolis, beeswax — any honeybee products or foods that are put on this earth to keep us in good health. Not foods loaded with artificial ingredients. All God-given food that is put on this earth for our bodies is perfect until man in his folly, from the time he was cast from the Garden — has tried to improve the produce of the earth. But man's efforts have not been a match for God's divine provision of the bee and its pollen, and its source of food.

It makes a great difference to your heart, mind, body and soul as to how long a life and how healthy a life you will live by what you eat and drink.

No matter how much exercise, taking pills of every kind, jogging, lying on a couch telling your troubles to someone or whatever your grandparents had will not build your body or give you a healthy body. It all depends on what "you" eat and drink.

Bee pollen can help correct the body chemistry and allows the body to rid itself of all unhealthy conditions.

Honeybee pollen contains all 22 essential nutrients needed for perfect health.

Ambrosia is simply a combination of unadulterated honey and bee bread — (bee bread is bee pollen).

Bee pollen is used in correcting errors of metabolism that is involved in unhealthy weight gain plus it is nature's method of weight control.

Bee pollen immunization can be achieved by incorporating pollen in food and drink.

They say you cannot cure a cold — you can keep from getting a cold by taking pollen and eating honey in your regular diet.

Bee pollen does relieve allergy symptoms.

The three highest foods having B_6 — methionine ratio are bananas, with a ratio of 40, carrots with a ratio of 15, and onions with a ratio of 10. Honeybee pollen has a ratio of 400-1, proving once again the miraculous properties of honeybee pollen as well as all honeybee products.

Athletes, as well as health-minded people and beekeepers, all over the world have known the health giving priorities of honeybee products since the beginning.

The number of minerals included in the functioning of the body is one of the wonders of life. Except for silver and gold, practically all minerals are in action.

To replenish minerals that the body needs, take 500 mg. of honeybee pollen a day. One easy and enjoyable way is to crumble a 500 mg. tablet of bee pollen in your mixer, add 1 banana, 1 orange, ½ pint of skim milk and 1 Tablespoon of pure honey.

Your entire body runs on "energy" all day and night. When you get tired and nervous, it's a sign your body needs energy. Bee pollen is loaded with all things the body needs for energy and well-being. Carry bee pollen tablets with you — when you get tired and nervous is the time to try 5 mg. of bee pollen.

Bee pollen is the most valuable ally in any program to partake fully of the elixir of human life.

POLLEN: THE BALANCED FOOD

"Only airborne pollens bring allergies - bees on gathering pollen add nectar and saliva which ensures grains are completely safe."

Health and happiness are normal states of mankind.

The healthiest people eat a well balanced diet of well balanced natural foods. Eat only when you are hungry and eat **when** you are hungry.

The composition and nutritional value of pollen and honey are the most perfectly balanced and minute form of any food known to mankind.

Pollen restores "balance" and normality to many who have lost it for so long that they have forgotten what true health is really like.

The real authorities know it is much more important to have a balanced and widely spread supplementation of vitamins than to take large quantities of a single one, except where there is a special need. Pollen clearly provides a good fundamental vitamin supplementation when used in addition to a normal, healthy diet. The minerals identified in pollen include sodium, potassium, magnesium, calcium, aluminum, iron, copper, zinc, manganese, lead, silica, phosphorus, chlorine, and sulphur. Enzymes already discovered in pollen are amylase, catalase, cozymase, cytochrome systems, diapase, diastase, lactic acid, dehydrogenase, pectase, phosphatase, saccharase and succinic dehydrogenase.

No one is claiming a freedom from colds or influenza, but all of the scientific tests on colds and influenza, none have been proven better for prevention than the use of honey and pollen in your daily diet.

Pollen is the richest source yet revealed of vitamins, minerals, proteins, amino acids, hormones, enzymes and fats. Pollen is essential for all plant life and garnered by honeybees, making it a perfect food and preventive medicine.

Pollen is so perfectly balanced that it is a complete survival food, as well as being the most potent food known to man. In order to be a survival food, you would only need to add a roughage and water.

In recent research into pollen, discoveries are still being made about the wonderful therapeutic value of and how remarkable this pollen and honey is. It has been found that pollen is so rich in protein that a cupful contains as much protein as there is in a quarter pound of beef.

Pollen researchers and micro-nutritionists are finding out what the honeybee and "preventive medicine" doctors have known for generations, that the precious food substances and micro-nutrients in the pollen grain is one of the greatest of all preventive medicines. Extensive medical trials in all parts of the world convince more and more orthodox doctors that the health, virility and vitality of the human body depends not just on the basic food ingredients - proteins, fats, carbohydrates, minerals and vitamins - but also on minute quantities of biologically necessary elements which are found in pollen and no matter how advanced the science or medical world may have become, they cannot compare with the natural creations that were put here for our bodies.

Our bodies may survive on artificial coloring, man-made vitamins, foods from chemicals, etc., but not for long, or not without difficulty with your mind and body. If your body becomes sick, your mind will be sick. You don't catch diseases, you eat diseases, so you are what you eat. The mind is like the stomach — it's not how much you put in it, but how much it digests.

Another very important finding is that the pollen, to be effective, had to be natural and not chemically treated. The public is beginning to realize more and more that the "whole" of nature and not just the refined and isolated parts of our food and drink is the healthiest for our minds and bodies.

For those who argue that man is nothing but a mixture of chemicals, and that a few man-made chemicals will not hurt you — If we are a mixture of chemicals, those chemicals are produced in forms which constitute the basic building blocks of nature. If we use our common sense, we know we do not have to have a PhD to know we are far better off if we feed and treat our minds and bodies with "natural substances" that are harmoniously balanced with nature as these do not produce undesirable side affects.

We are not decrying the magnificent advances in medicine or in the use of quite artificial materials to control disease, but that many of these stop-gaps in man's medical progress will be replaced by natural substances.

At least you can try pollen and honey as a preventive. Some doctors are so concerned with modern drugs that the healing powers of simple remedies are lost in the passing of time.

Throughout the ages, honey has always been accepted as the finest food available to man. In every land inhabited by the honeybee, the same beliefs existed regarding the miraculous powers of honey as both a food and medicine. Writers and "learned men", since the beginning of civilization have known this nectar to be the most wonderful food and storehouse of medical goodness.

For centuries the power of natural food was accepted. Scientists were able to prove conclusively that honey is not only a perfect food but also a giant germ killer in which bacteria simply cannot exist. In this ever-changing world, honey is as good today as ever, being one of the few things that the atomic age has passed by.

The composition of this amazing food is no mystery. Man today can answer nearly all the questions about honey, but in spite of this, some 2 percent of its contents still remain a mystery and defies identification. There is no food or preventive medicine known that is older or has proven itself more than honey has.

Some doctors, chemists, nutrition experts and government agencies, all of whom believe in theory, not facts, are so concerned with modern drugs, chemicals, man-made foods, that the healing powers of simple remedies and God-given foods and drinks are lost in the passing of time, but not to the experienced, who believe in facts, not theory, probability or practice, which prove nothing. They say medicine has come a long way. The culpable disregard of honey and pollen by these kind of men who seem to believe that the human body can continue to survive entirely on theory and man-made chemicals is a grave and lamentable error of the present generation and a sad reflection on its own intelligence.

There is no food or preventive medicine older or that has proven itself more than honey and pollen.

The pollen grain is protected by two durable coats. The outer waxy exine is made of sporopollen which can resist most acids and temperatures as high as 300° centigrade. Beneath this is the fragile inner wall, the intine, which surrounds and protects the nuclei and the reserves of starch and oil. The pollen grain is practically indestructible.

NOTES

GATHERING AND STORING HONEYBEE POLLEN

The tongue and mandibles are used in licking and biting the anthers with the result that pollen grains stick to the mouthparts and become thoroughly moistened. A considerable amount of pollen is dislodged from the anthers, and adheres to the hairy legs and body. The branched hairs of the bee are suited to retaining the pollen which is dry and powdery.

After the bee has crawled over a few flowers, she begins to brush the pollen from her head, body, and forward appendages and to transfer it to the posterior pair of legs. The wet pollen is removed from the mouthparts by the forelegs. The dry pollen clinging to the hairs of the head region also is removed by the forelegs, and added to the pollen moistened by the mouth.

Much sticky pollen is now assembled on the inner faces of the broad tarsal segments of the second pair of legs.

Pollen is transferred to the pollen baskets.

Each new addition of pollen is pushed against the last and, simultaneously, the masses of pollen on both legs grow upward, a very small amount being added at each stroke.

Finally, each leg is loaded with a mass of pollen, held in place by the long recurved hairs of the elevated margins of the tibiae. If the loads are very large, these hairs are pushed outward and become partly embedded in the pollen, allowing the mass to project beyond the margins of the tibiae.

When the bee is loaded, she returns to the hive. Some walk normally over the combs, while others appear to be greatly agitated, performing the characteristic "dance" which communicates to the other fielders the existence of a source of pollen. Many pollen-bearing bees "solicit" food from other workers or take it from the cells.

The bee moistens the pellets with its tongue.

Foraging trips of pollen gatherers are considerably shorter than those of nectar gatherers. The number of flowers visited by pollen gatherers, the time spent in making a load, the number of trips per day, and the weight of pollen loads is variable, depending on the species and condition of the flowers, temperature, wind velocity, relative humidity, and possibly other factors. To make a load of pollen, a bee visited 84 flowers of pear trees, and 100 flowers of dandelion. To make a full load of pollen a bee may spend 6 to 10 minutes or as much as 187 minutes. The number of trips per day may be 6 to 8 and up to 47, the average probably being about 10 trips per day. Using pollen traps, during good gathering weather, between 50 to 54 thousand bees bring pollen into the hive daily. The weight of the pollen loads ranged from 12 mg. for elm to 29 mg. for hard maple fresh weight, or 8.4 to 21.4 mg. dry weight. Marked differences are found in the amount and the character of pollen brought to the hive by various colonies of the same apiary. In the spring pollen is collected at temperatures as low as 8° to 11° C. High relative humidity decreases pollen collection.

The quantity of pollen carried on the body of the honeybee is larger than that of any other hairy insect.

Chemical Analysis of Honeybee Pollen
Honeybee Pollen and Honey are the Only Food on Earth
Containing All 22 Nutrients Needed by Mankind
For Complete and Perfect Health.

VITAMINS
Provitamin A (carotenoids)
 5-9 mg %
Vitamin B1 (thiamine)
 9.2 micrograms %
Vitamin B2 (riboflavin)
Vitamin B3 (niacin)
Vitamin B5 (panothenic acid)
Vitamin B6 (pyridoxine)
 5 micrograms %
Vitamin B12 (cyamoco balamin)
Vitamin C (ascorbic acid)
Vitamin D - Vitamin E
Vitamin H (biotin)
Vitamin K, Choline, Inositol
Folic Acid, 5 micrograms %
Pantothenic acid
 20-50 micrograms/gram
Rutin, 16 milligrams %
Rutin in beehive pollen 13%
Vitamin PP (nicotinicamide)

MINERALS
Calcium, 1-15% of ash
Phosphorus, 1-20% of ash
Iron, 1-12% of ash
 .01-1.3% of fresh pollen
 .6-7.1 mg % of air dried
Copper, .05-.08% of ash
 1.1-2, 1 mg % of fresh
Potassium, 20-45% of ash
Magnesium, 1-12% of ash
Manganese, 1.4% of ash, 75 mg
 75 mg %
Silicoa, 2-10% of ash
Sulphur, 1% of ash
Sodium - Titanium - Zinc
Iodine - Chlorine
Boron - Molydbenum

ENZYMES & COENZYMES
Disstase   Phosphatase
Amylase   Catalase
Saccharase   Diaphorase
Pectase   Cozymase
Cytochrome systems
Lactic dehydrogenase
Succinic dehydrogenase
Note: The cozymase in mixed fresh
pollen runs about 0.5-1.0 milligram
per gram, comparable to the amounts
in yeast. (Bee pollen contains all
known enzymes & co-enzymes and
probably all that will be known
in the future.)

FATS & OILS — 5%
Fatty acid (may be 5.8%)
Hexadecanol may be 0.14% of
pollen by weight. Alpha-amino
butyric acid is present in pollen
fat. Unsaponifiable fraction of
pollen may be 2.6% by weight.

FATTY ACIDS (Conifer Pollen)
Total list identified are:
Caproic (C-6) - Caprylic (C-8)
Capric (C-10) - Lauric (C-12)
Myristic (C-14) - Palmitic (C-16)
Palmitoleic (C-15) one double bond
Uncowa - Stearic (C-18)
Oleic (C-18) one double bond
Linoleic (C-18) two double bonds
Arachidic (C-20) - Benemic (C-22)
Limolenic (C-18 three double bonds)
Eicosanoic (C-20 one double bond)
Brucic (C-22 one double bond)
*Pseudotduga dry pollen contains
0.76-0.89% fatty acid. Major are:
Oleic, Palmitic, Linoleic, *Pinus
dry pollen contains: 1.25-1.33%
fatty acid based on dry weight of
pollen. Major are: Linolenic,
Oleic, Stearic.

PIGMENTS
Xannmepayll, (20-150 micrograms
 per gram.)
Carotates (50-150 micrograms per
 gram.) Alpha & Beta Carotene
Chlorophyll, in hand-collected, but
 not bee-collected pollens.
Ammpcyamin, in hand-collected.
 but not bee-collected pollens.
Crocetin, Zaaxanmin, Lycopene

WATER - 3-20% of fresh pollen

MISCELLANEOUS
Waxes, Resins, Steroids, Growth
Factors, Growth Inhibitors, Vernine
Guanine, Xanthine, Hypoxanthine,
Nuclein, Amines, Lecithin, Glucoside of
Isorhanetin, Glycosides of
Quercetir, Selenium, Nucleic acids
flavonoids, phenolic acids, tarpenes
& many other yet undefined nutrients.

PROTEINS, GLOBULINS, PEPTONES, AMINO ACIDS
★ 7-35%, average 20%: 40-50%
may be free amino acids: 10-13%
consists of amino acids in dry pollen.
★ 35 grams of pollen per day can sat-
isfy the protein requirements of man.
25 grams of pollen per day can sustain man because it contains 6.35
grams as indicated by Rose, plus
other amino acids. ★ Pollen contains
the same number of amino acids,
but vary greatly in quantity of each.
Tryptophan 1.6% - Leucine 5.6%
Lysine 5.7% - Isoleucine 4.7%
Methionine 1.7% - Cystine 0.6%
Thresonine 4.6% - Arginine 4.7%
Phenylalanine 3.5% - Histidine 1.5%
Valine 6.0% - Glutamic acid 9.1%
Tyrosine - Glycine - Serine
Proline - Alanine - Aspartic acid
Hydroxyproline - Butyric Acid

CARBOHYDRATES
Gums - Pentosans - Cellulose
Sporonine (7-57% of pollen of
 various species; 28% in bee-
 collected, 57% hand collected)
Starch (0-22% of pollen)
Total sugars (30-40%)
 Sucrose or cane sugar
 Levulose or fruit sugar/fructose
 Glucose or grape sugar
 Reducing sugars (0.1-19%)
★ Bee-collected: Non-reducing sugar
 2.71%. Reducing, 18.82-41.21%
 Mean, 25.71%
★ Hand-collected: Values reversed
 Air-dried: Non-reducing 0-9%
 Reducing sugars, 20-40%
★ Hand-collected: Non-reducing,
 up to 22%. Reducing, 0-7.5%

TWENTY-EIGHT MINERALS are found in the human body. Fourteen are vital,
essential elements present in such small amounts that they are called "micro-
nutrients". HONEYBEE POLLEN CONTAINS ALL 28 MINERALS.

Nucleosides	Guanine	Hexodecanol
Auxins	Xanthine	Alpha-Amino-Butyric Acid
Brassins	Hypoxalthine	Monoglycerides
Gibberellins	Crocetin	Diglycerides
Kinins	Zeaxanthin	Triglycerides
Vernine	Lycopene	Peutosaus

★ A NUTRIENT is a MOLECULE you must have, but the body cannot manufacture.
 You have to ingest (eat) it. If you don't have it, at first you will not feel well. If you
 don't get it for a longer time, you will begin to feel sick. If you don't get it for TOO
 long a time, you are probably going to die.

Each Ounce of Honeybee Pollen Contains:
JUST 28 CALORIES
ONLY 7 GRAMS OF CARBOHYDRATE
PLUS 15% LECITHIN, the substance that burns away fat.
AND 25% IS PURE PROTEIN!

3 teaspoons =1 Tablespoon.   2 Tablespoons = 1 ounce.

POLLEN AND ALLERGIES

We inherit the tendency to develop an allergy but not the allergy itself. In hayfever, the irritant is usually pollen from ragweed, although tree and grass pollen can affect some people all year. Pollen is the yellow powdery dust we see in great concentrations on flowers. The pollen grains stick to the furry legs of bees collecting nectar.

What becomes difficult for many doctors to comprehend is why people eat substances to which they are known to be allergic. The concept behind such a seemingly harmful act is that putting amounts of the irritants into the body, one induces the system to build defense or immunities against the foreign substance. Thus, pollen, has been proven by so many to build up immunity to hayfever and to relieve victims of the symptoms. Literature on allergies in general is abundant; the question of pollen as a cure for hayfever and other allergies is almost ignored.

P. M. Winter, PhD, of Kingston, Pa. and many, many others, consider pollen to be the best cure. Some doctors say the pollens must be injected. However, pollen goes quickly to the bloodstream and goes through the bloodstream at a trickle, so some of the pollen protein does not go through the digestive process but is assimilated unchanged into the bloodstream.

Drugs can have bad side-effects and even injections, the major form of therapy, have not always proven effective.

If your doctor has not been able to help you after testing and treatment, ask your doctor if under his supervision you can try taking by mouth, small amounts every day of the allergenic material.

Patch and rasp allergy test for food sensitivity is of no value in diagnosing which foods may be causing your trouble. These tests only measure the reactivity of the skin to the particular food extract.

Hayfever and other allergies caused from the pollen in the air, can be stopped by taking honeybee pollen.

NOTES

STATEMENTS BY DOCTORS

Dr. Carlson Wade, in "About Pollen", states: "Bee pollen contains a gonadotrophic hormone similar to the pituitary hormone, gonadotrophin, which functions as a sex gland stimulant. The healing rejuvenating and disease-fighting effects of this total nutrient are hard to believe, yet are fully documented. Aging, digestive upsets, prostrate diseases, sore throats, acne, fatigue, sexual problems, allergies, and a host of other problems have been successfully treated by the use of bee pollen."

Dr. Carlton Fredericks: "Honeybee pollen is the only super perfect food on this earth." This statement has been proven so many times in the laboratories around the world by a chemical analyst that it is not subject to debate nor challenge.

Dr. Bernard C. Jensen: "Much has been said about pollen helping glands in the body. All experiments on animals show that it prolongs life and helps keep glands in good order."

Dr. Sigmund Schmidt: "Eat pollen. Pollen contains all the essential elements, vitamins and minerals, for healthy tissue and therefore could be a cancer preventative."

Leo Conway, M.D., 1972, of Denver, Colorado, had treated in excess of 60,000 documental and verified cases of allergies with pollen and says, "I believe this pollen immunization can be achieved by incorporating pollen in a food, because the only need for proof is the effectiveness of the commodity through the digestive tract."

Dr. Kilmer McCully, M.D., Professor of Pathology at Harvard Medical School: "The three highest foods having the B_6 methionine ratio are bananas with a ratio of 10, carrots with a ratio of 15, and onions with a ratio of 10. Honeybee pollen has a ratio of 400 to 1." Proving again the miraculous properties of honeybee pollen as superior to any other food.

We could go on and on.

NOTES

"HIPPOCRATES"

The Hippocratic Oath, named for him, gave the medical profession a sense of duty to mankind, which is never entirely lost. These people have been their own nutritionalists and have brought up their families to be healthy by going by facts and not theory.

Honeybee pollen is manufactured by the honeybee in this manner . . . The honeybee travels from flower to flower, gathering pollen on its hairy body. All things on earth are male and female; the bee travels from the female stigma of one flower to the anther of another flower. The pollen falls from one flower to the other pollinating the flowers and all of our crops. Without the honeybee pollination we could not have the crops we now have.

Crops are pollinated by rain, wind, butterflies, birds and honeybees; but the honeybees are the only pollinators we can depend on.

As the honeybee gathers pollen on its hairy body it takes the pollen in its mouth and chews it, mixing the pollen with its saliva. The honeybee's saliva has enzymes that only the honeybee produces. When they mix the pollen and saliva, they pack it in the pockets on their hind legs and return to the hive where they store the pollen for their use.

The beekeeper puts a pollen trap on the beehives. As the honeybees enter the hive they are forced to enter through the pollen trap. The bee drops part of the pollen off their hind legs into the traps . . .. this is the bee pollen used for quick energy, preventive medicine and pure food. Also if you eat this kind of pollen there is nothing better to have running through your bloodstream to ward off pollen allergies — hay fever, etc.

NOTES

POLLEN TRAPS

The most important aspect of pollen production is the correct pollen trap. Most traps are complicated, inefficient and not practical. There are many traps available but their claims are meaningless.

A trap must be a simple one to operate; not with pull-out screen, flip-up trap screen, turn-over drawers and upside-down gimmicks that let in dirt and windblown objects as a lot of expensive, fancy traps. The fancier the trap the less efficient they are.

You need at least 11"x11" space for a good honey crop. Never get a trap with too small a screen area or you will reduce your honey production. Also have at least ⅜"x½" drone escape hole.

If pollen is as much as 25% moisture it will go rancid in 4 to 5 days if not refrigerated.

Pollen absolutely has proven to be nothing short of a miracle food. A number of athletes and professional sports teams around the world have used pollen to gain the competitive edge.

If you want to be a healthy, energetic person . . . try bee pollen.

The inclusion of pollen as a product of U.S. beehives can be a benefit to the bee industry.

Pollen has an exceptionally high vitamin content, especially the water-soluable vitamins.

Sight and smell enables the bees to locate sources of pollen and nectar. The behavior during flower visitation depends upon whether the forager collects pollen only, nectar only, or both and differs also with the type and size of the flower.

NOTES

POLLEN AND HONEY
NATURE'S GOLDEN TREASURE

Bees are skillful little creatures who serve one another and unwittingly, they serve us. The industrious workers make three times as much honey as the hive needs, which enables us to enjoy this nutritious food.

Even more important, bees pollinate flowers. As they gather nectar, pollen grains from neighboring blossoms are brushed off their bodies and legs. Without this "accident" of nature, we would not have many of the fruits and flowers we take for granted today.

Honey, being rich in real fructose, is a vital food. In short time, one hive multiplies to several. The beekeeper has honey for his family and plenty left to take to market. He can count on a regular income from the sale of honey and beeswax.

As the number of bees in an area increases, so will the quality of local fruits and vegetables. And cross-pollination will brighten the fields with flowers.

In many ways, bees help make a community a better place to live.

California leads the nation in three important enterprises of the beekeeping industry. They are pollination service, the production of queens and package bees, and of course, honey and beeswax.

Honeybees are used to pollinate crops worth over a billion dollars per year in the United States. Over ninety crops are dependent upon the honeybee for pollination. It is easy to understand why the honeybee is so important to the nation's agriculture economy.

Many industries benefit from beekeeping. For example, beeswax is used for making beautiful candles that burn slowly and do not drip. Beeswax is also used in many types of cosmetics, floor waxes, furniture polish and paints.

In the Spring of the year many beekeepers introduce new queens to their hives. These queens are produced, packaged, and shipped to their new homes by queen breeders. This is another big business and increases the revenue of printing firms, advertising, packaging and shipping companies, as well as the post offices throughout the nation. Queen bees and their escorts often arrive at their destinations via the U.S. Mail.

For many centuries, honey was the world's major source of sugar and sweets in the diet, as it still is in many countries. The United States has other sources of sugar, however, notably beets and sugar cane which are used to make refined sugar.

With increased attention being given to the health problems of our nation, bee pollen and honey are enjoying increasingly wider usage. More and more people who enjoy sweets in their diet do not care to use refined white sugar, which is devoid of healthful properties.

Second to none in taste appeal and variety of uses, honey is a natural, unrefined food, unique because it is the only unmanufactured sweet available in commercial quantities. In addition to its sugars, honey contains a considerable number of minerals, seven members of the B Vitamin complex, ascorbic acid (Vitamin C), dextrins, plant pigments, amino acids and other organic acids, traces of protein, esters and other aromatic compounds, and several enzymes.

Since seventy-five to eighty percent of its composition is sugars, honey has an energy-producing value second to few foods. The important thing is the way in which honey provides energy. Cane and beet sugars must be broken down into simpler sugars by digestive juices before they can be absorbed into the bloodstream and assimilated into the tissues. These resulting simple sugars, dextrose and levulose, make up almost the entire sugar content of honey. As a result, little digestion is necessary, and absorption takes place quickly.

This is one reason why honey is extremely popular among athletes of all types. It produces virtually instant energy without putting any strain on the digestive system. Football, basketball, baseball and hockey players use pollen and honey before and during competition for quick energy, as do all types of sportsmen.

Americans consume an average of 285,000,000 pounds of honey every year. That means a fantastic amount of work for millions of little honeybees. There is no harder worker anywhere in the world. The average life of a worker bee is about six weeks, after which the bee, its wings literally worn to shreds from flying, simply dies from exhaustion. It takes 556 worker bees flying 35,584 miles (1⅓ the distance around the world) to produce one pound of honey! If honey were to be priced according to the effort that goes into its production, none of us could afford to enjoy it.

The indiscriminate use of insecticides to control insect pests has killed off most wild bees which formerly pollinated crops across the nation.

Honeybee Pollen for Health

Here are just a few ways to use Honeybee Pollen in some healthy drinks. Bee pollen may be added to any of your favorite recipes such as breads, cakes, cookies and salad dressings. The bee pollen pods may be crushed and used or when added to liquid they will blend or melt. Bee pollen is healthy and delicious in these drinks — 500 mg. bee pollen equals 1 teaspoon in these recipes.

Put Some Juice In Your Life Now

Instead of refreshing yourself with super-sweet high-calorie carbonated beverages, try fresh fruit juices. They do contain calories, but not as many as carbonated beverages; they also contain important nutrients. They also impart an incomparable fresh taste. For variety, mix two or three more different juices together with 500 mg. honeybee pollen pods.

For example: Combine a quart of apple juice with 1 cup grape juice and ¼ cup lime juice. Pour over ice in 6 tall glasses.

For an instant breakfast: Blend 1 cup apple juice with 1 banana, 1 egg, and 1 tablespoon honey in an electric blender.

HEALTHY BLENDER DRINK

1 ripe banana
2 prunes, cooked and pitted
1 T. all-bran cereal (shredded form)
1 T. wheat germ nuggets
½ t. lemon juice
½ C. orange juice
1 T. honey
500 mg. pollen pods

Place in blender all ingredients; blend until smooth and creamy. Makes about 10 ounces—a breakfast beverage for two, a light meal for one.

NOTE: Try substituting one-half cup milk for the orange juice and add one tablespoon natural peanut butter (with or without the prunes). Be creative!

HONEY PUNCH

Honey Syrup:
4 C. boiling water
1½-2 C. mild-flavored honey
500 mg. pollen pods

Honey syrup is made by blending together 4 cups boiling water with 1½ to 2 cups of mild-flavored honey. Keep in refrigerator in covered container.

Punch:
½ C. honey syrup
½ C. fresh lemon juice
1½ C. apple juice
1½ C. fresh orange juice
4 C. ice water
Lemon slices

Combine all ingredients except lemon slices. Stir until well blended. Chill. Serve in tall ice filled glasses. Makes 2 quarts.

CARROT-PINEAPPLE COCKTAIL

2 C. unsweetened pineapple
  juice
2 medium carrots, washed and
  cut into 1-inch pieces
1 slice lemon, ¼" thick
1 C. ice, crushed
2 t. honey
500 mg. bee pollen pods

In an electric blender combine the pineapple juice, carrot pieces and lemon; blend until carrot is liquefied. Remove cover from blender and add the crushed ice. Cover and continue to blend until ice is liquefied. Sweeten with honey. Serve in cocktail glasses garnished with a slice of orange. This recipe makes approximately three cups.

GOLD STRIKE

1 egg
1 C. orange juice
1 T. honey
100 or 500 mg. bee pollen pods

Combine egg, juice, honey, and pollen pods. Beat or shake until well blended. Pour into tall glass. Serve immediately. Makes one serving.

VARIATION: Just before serving add a scoop of orange or lemon sherbet.

This drink makes an excellent breakfast drink for those in a hurry on warm summer mornings.

PEACHY-HONEY FLOAT

2 C. crushed fresh peaches
½ C. honey
1 qt. milk
½ t. almond extract
1 qt. vanilla or cherry ice cream
  (homemade)

Combine fresh peaches and honey in container of blender. Add half of milk; beat and blend. Add balance of milk, almond extract and half of ice cream. Beat until smooth. Pour into tall glasses, top with balance of ice cream. Yield: 6 servings.

HONEY EGGNOG

2 eggs, beaten
2 T. honey
2 C. cold milk
½ t. vanilla
Dash of ground nutmeg
500 mg. bee pollen pods

Combine eggs with honey and pollen pods and mix well. Beat in milk and vanilla. Sprinkle lightly with nutmeg. Makes 2 servings.

HONEY ORANGE EGGNOG

500 mg. honeybee pollen
6 eggs, separated
5 T. honey
3 C. heavy cream, divided
2 C. milk
1 C. orange juice
Grated rind of one orange
Ground nutmeg

Beat egg yolks with three tablespoons of honey. Add 1 cup cream and milk and continue beating. Add the orange juice slowly to prevent curdling the milk. Whip remaining two cups of cream; fold into orange mixture. Beat the egg whites until soft peaks form, then add two tablespoons of honey. Beat again until honey is mixed into egg whites. Fold egg whites into the eggnog. Garnish with the grated orange rind and nutmeg.

HONEY HEALTH PUNCH

2 - 500 mg. honeybee pollen pods
6 C. apple cider or juice
1 cinnamon stick
¼ t. ground nutmeg
¼ C. honey
3 T. lemon juice
1 t. grated lemon peel
1 (18-oz.) can unsweetened
   pineapple juice (2½ C.)
Additional cinnamon sticks

In large pan, heat cider and cinnamon stick to boiling; reduce heat. Cover; simmer for 5 minutes. Uncover and stir in remaining ingredients except additional cinnamon sticks and simmer 5 minutes longer. Use cinnamon sticks as individual stirrers. Makes 16 (½ cup) servings.

HOT HONEY CIDER

2 - 500 mg. honeybee pollen pods
1 medium orange, cut into 5
   slices
2 qts. apple cider
½ C. honey
1 t. whole allspice
16 whole cloves
10 cinnamon sticks

Cut 2 orange slices into quarters, and set aside. Combine remaining orange slices, cider, honey, allspice, cloves and 2 cinnamon sticks in a 3-quart saucepan; bring to a boil. Reduce heat, and simmer 15 minutes. Stir well; pour into serving mugs; garnish each with a quarter of an orange slice and a cinnamon stick. Yield: 8 servings.

STRAWBERRY-HONEY FLOAT

500 mg. honeybee pollen pods
1 qt. milk, chilled
6 T. honey
2 C. crushed fresh strawberries
½ t. almond extract
1 qt. homemade vanilla ice
    cream

Combine milk, honey, straw-berries, almond extract and one pint ice cream. Beat with rotary beater until blended. Pour into tall glasses and garnish with scoops of ice cream. Makes 6-8 servings.

CRANBERRY FIZZ

For 1 serving:
¾ C. club soda
¼ C. cranberry juice
1 egg
2 t. honey
500 mg. honeybee pollen pods

Measure ingredients into 5-cup blender container. Blend well. Pour into glass, pitcher or small punch bowl. Serve immediately.

APPLE SIDEUP

¾ C. apple juice, chilled
¼ cup milk, chilled
¼ t. ground cinnamon
1 t. honey
1 egg
500 mg. honeybee pollen pods

Combine all ingredients in shaker, blender or mixing bowl. Shake or beat to a froth or blend well at low speed. Pour into tall glass. Serve immediately. Makes one serving.

BEE POLLEN AND HONEY PICKERUPPER

500 mg. honeybee pollen pods
1 egg
1 C. skim milk
1 C. fresh fruit
1 T. honey
2 T. wheat germ
3 ice cubes

Put first 5 ingredients into blender. Cover and process at LIQUEFY. Add ice cubes one at a time. Con-tinue to process until smooth. Yield: 2 servings.

ORANGE HONEYADE

500 mg. honeybee pollen pods
2 C. orange juice
½ C. lemon juice
½ C. honey
1 C. water.

Combine ingredients and stir well to dissolve honey. Pour over cracked ice in tall glasses. Gar-nish with orange slices or cherries.

HOW TO MAKE HONEY ROOT BEER
IN 24 HOURS

One gallon glass jug with tight screw-cap to prevent leaking or loss of
 carbonation
One and three-fourths cups of mild honey
One-half of a three-ounce bottle of root beer extract
One-fourth teaspoon of dry yeast

Place 1¾ cups of mild honey, ½ 3-oz. bottle of root beer extract and ¼
teaspoon of dry yeast in the gallon jug, which should be ½ full of warm
water.

Shake the mixture thoroughly and then fill the bottle to within ½ inch
from the top with more warm water. Then rotate and turn the bottle to complete blending.

Lay the bottle on its side in a cool place for 24 hours, then refrigerate to
temperature desired, then pour and enjoy the healthiest root beer.

You can also make a delicious root beer float by adding one or two dip-
pers of honey ice cream to a glass of root beer.

If you have trouble finding root beer extract, you may write to Mr. Mike
Brady, Sales Service, Fountain Division, Crush International, Inc., 2001
Main Street, Evanston, Illinois 60202. Mr. Brady will ship Hires root beer
household extract in a minimum quantity of twelve 3-oz. bottles. Just the
right size for the recipe and enough to make 24 gallons of root beer. Price is
$8.25. 1½ ounces of root beer extract gives a good flavor. Sweet clover or
alfalfa honey is my favorite.

HONEYBEE POLLEN FOR SKIN CARE

Your skin can become younger looking and less vulnerable to wrinkles, smoother and healthier with the use of bee pollen. Not only does pollen help clear up acme conditions in the young, it is also of special benefit in rejuvenation for older skin.

Honeybee pollen will prevent premature aging of the cells and stimulate the growth of new skin tissue. It smooths away wrinkles and stimulates the life giving blood supply to all cells because it contains high concentration of nucleicacids RNA and DNA. These substances penetrate the surface of the skin when a lotion containing bee pollen is used and they nourish the cells and tissue beneath. They act as a moisturizer for dry skin subject to wrinkling. Also helps smooth furrows and creases.

Improve the look and feel of your skin with the use of honeybee pollen in all natural beauty aids. Here are a few I think are very good. These recipes have been used in our family for generations.

HONEYBEE POLLEN FACIAL MASK

500 mg. bee pods equals 1 teaspoon.
Mix:
½ ripe avocado
1 Tablespoon honey
¼ cup whole milk
2 Tablespoons bee pollen pods

Have all ingredients at room temperature. Either in a jar by hand or blend in a blender all ingredients. Blend well. Now clean your face and throat thoroughly. With fingers apply this mask. Leave on for 20 minutes then rinse with warm water. Follow with a brisk patting of astringent made bee pollen. "See below."

HONEYBEE POLLEN ASTRINGENT

2 Tablespoons lemon juice
1 Tablespoon glycerin
1 Tablespoon 70 percent alcohol
1 Tablespoon bee pollen pods

Add just enough distilled witch hazel to fill 8-oz. cup. Stir vigorously with a spoon or in a blender. Refrigerate until tingly cool then apply with a cotton ball or finger tips to your face and throat. Leave on most of the day.

FACIAL CLEANSER

Beat the yolk of one egg until it is light and frothy. Add ½ cup milk, ½ mashed ripe avocado and 1 teaspoon bee pollen pods. A blender is handy here but if you don't have one, beat the mixture with a fork until you have a thin creamy lotion. Apply with cotton swabs or as you would any other cleanser.

You may use this deep cleanser after cleansing with soap and water. Since this formula is perishable, it is best to make it every few days and store in the refrigerator between uses.

MASK FOR OILY SKIN

Put the white of an egg, 1 teaspoon lemon juice, 2 teaspoons bee pollen pods, and ½ avocado into a blender or beat vigorously by hand. Wash face and neck and apply the mask. Relax for 20 minutes. Rinse with cool water. Follow with cold astringent.

BEAUTIFUL HAIR

Add a few drops of lemon juice and a teaspoon of bee pollen pods to your favorite shampoo. Use as usual. Your hair will be left with a lovely shine.

HONEYBEE POLLEN SCRUB

Once a week your face should be given a good scrubbing to get rid of the dull muddy look that comes from not quite removing all the dirt and makeup.

Mix together the juice of one lemon and the white of one egg. Add 500 mg. bee pollen, add oatmeal until you have a soft paste then allow to set a few minutes until the moisture is absorbed. Apply to your face avoiding the eye areas. Rub in gently with a tender scrubbing motion. Let dry 10 minutes. Rinse with warm water. Follow with very cold water.

HONEYBEE POLLEN FOR DANDRUFF

In a bottle shake 1 cup white vinegar, 1 cup water, 2 Tablespoons bee pollen pods. Dab this solution using cotton pads onto your scalp before shampooing. Ingredients in this dandruff lotion will help loosen scalp debris and dandruff and will help cleanse your scalp and leave your hair squeaky clean.

OATMEAL FACIAL PACK

⅓ c. finely ground oatmeal
3 t. honey, enough to make a
  smooth paste

1 t. rose water (or orange
  flower water)
500 mg. bee pollen pods

Blend oatmeal with honeybee pollen until well mixed. If too thick and unmanageable, add a little rose water or orange flower water. Spread over clean face with the exception of your eyes, and leave it on for about one-half hour. Relax while it is on if you can. Remove with soft washcloth and warm water. A good astringent should be applied to tone the skin.

Variations:

HONEY BRAN PACK

Omit oatmeal and add one-half cup bran.

HONEY CORNMEAL PACK

Omit oatmeal and add ⅓ cup cornmeal.

HONEY PASTE PACK

Omit oatmeal and add one teaspoon fine white flour.

ALMOND HONEY FACE LOTION

1 T. sweet almond oil

2 T. honey
500 mg. bee pollen pods

Blend together. This lotion should be used after the skin has been thoroughly cleansed. It should be permitted to remain on the skin about ½ hour. Then remove it with a soft cloth and tepid water. Apply milk astringent to close the pores and tone the skin.

SOFTENING HONEY FACIAL

¼ t. apple cider vinegar
  or lemon juice

1 T. honey
500 mg. bee pollen pods

Beat well and spread liberally over the face. Leave on for 15 minutes and rinse off in warm water. Pat dry.

CLEOPATRA'S FACIAL BALM

1 t. honey
1 egg white

1 t. milk
500 mg. bee pollen pods

Beat well and apply to clean face and neck. Leave on ½ hour. When it feels dry and brittle, wash it off with lukewarm water. Splash on cold water. You will feel your face and neck tingle. Cleopatra used this formula on her entire body to keep her skin beautiful and soft.

NOTES

Honey

EAT HONEY TO LOSE WEIGHT

The scientific reason for losing weight by eating honey and bee pollen.
When you eat honey or bee pollen it goes into the bloodstream, causes rapid combustion and consumes fats which speed up the burning of fat, and continues through the bloodstream at a trinkle and stimulates the heart without harmful after-effects.

NOTES

GATHERING NECTAR

In order to make pure honey the honeybee, as it travels from flower to flower, puts its "provis" or long tongue down into the flowers and sucks up "nectar". The tongue is built like a straw. When it sucks up the nectar it goes into the honeybee's honey sack where it mixes with the enzymes (that only the honeybee has) in the honey sack. After the honeybee returns to the hive it passes the nectar from bee to bee and they store the nectar in the honeycomb.

The nectar has more water content than is needed in the honey at this time so the bees fan their wings and evaporate the extra water content from the nectar until it reaches approximately 18 percent. This is known as "aging the nectar". Then and then only is the honey the finished unadulterated honey. No one can improve on honey . . . except the honeybee.

Everyone knows that bees make honey, yet no one, not even the scientists, knows exactly how they make it. In fact, mankind can't reproduce synthetically the product created in the wild. The honey making is the work of specialists known as house bees. It is their job to wait at home and receive the nectar from the field bees. By means of evaporation they reduce its moisture content until only the concentrated essence remains. The eventual conversion to honey is accomplished through a mysterious digestive process. When the golden liquid is finally of the right consistency, it's sealed in a honeycomb.

Nectar is a sweet liquid secreted by plant nectaries usually located within the flowers, but in some species they are situated elsewhere. Nectar is the reward "offered" to bees in return for their indispensable services in cross-pollination.

She alights upon any convenient part of the plant that will support her weight. Upon alighting, the proboscis is brought forward from its inactive position beneath the "chin" and is inserted into that part of the flower where nectar accumulates.

Bees are guided to nectaries by the difference in the odor of the nectaries and of the rest of the flower. A bee cannot tell whether there is nectar in a given blossom without inserting her tongue. By this means, however, she very quickly determines its presence or absence. When nectar is found, the bee sucks nectar within reach of the tongue until all has been taken up. In case none is found, the proboscis is withdrawn immediately and she moves quickly to another flower or floret.

Several hundred visits may be necessary to obtain a load of nectar from small flowers such as those of sweet clover. Any species from a load may be gathered in less than 100 visits would be a highly desirable honey plant.

A nectar gatherer spends on an average from 27 to 45 minutes per trip depending on the nectar flow. Bees work from 106 to 150 minutes to get a load of nectar. The duration of a trip depends on the species of flowers visited, and on the condition of the honeyflow.

Large loads of nectar weigh on the average about 70 mg. or 85 percent of the weight of the bee which, in the case of Italians, was found to be approximately 82 mg. Average loads of nectar during a honeyflow weighed about 40 mg.

The forager loaded with nectar enters the hive and moves to a place among other workers on the comb. If the nectar flow is weak she walks about until she meets a house bee to which she gives part of her load. Occasionally she gives her entire load to a single house bee, but more often it is distributed among three or more. If the nectar source is bountiful, the loaded nectar gatherer usually "performs" the dance already referred to as a means of communication. At irregular intervals, the dancer pauses and "offers" a taste of the nectar to one or another of the nearby workers. But soon she meets a house bee to which she gives a considerable portion of her load. As they approach each other, the field bee opens her mandibles wide apart and forces a drop of nectar out over the upper surface of the proximal portion of her proboscis, the distal portion being folded back under the "chin." Assuming that the house bee approached is not already loaded to capacity, she stretches out her proboscis to full length and sips the nectar from between the mandibles of the forager. While the nectar is being transferred in this manner, the antennae of both bees are in continual motion, and those of one bee are constantly striking those of the other. At the same time, the house bee may be seen to stroke the "cheeks" of the field bee with her forefeet. This may further stimulate unloading behavior.

Upon disposing of her load, a nectar gatherer sometimes leaves for the field immediately, but usually she pauses long enough to secure a small amount of food. In any case her departure is immediately preceded by certain characteristic maneuvers. She first gives her proboscis a swipe between her forefeet, then rubs her eyes, and often cleans her antennae, then starts for the field. The entire process of disposing of her load often is accomplished in less time than it takes to describe it.

STORING AND RIPENING NECTAR

In the manufacture of honey from nectar, two distinct processes are involved: One brings about a chemical change in the sugar and the other results in a physical change whereby surplus water is eliminated. Once the honey is "ripe" it is sealed in cells with beeswax caps.

The sugar content of nectars is very small amounts of sucrose present in a few nectars. Due to the action of the enzyme invertase, sucrose is changed in the hive into two simple sugars, glucose and fructose. When the house bee has received the field bee's load, she moves about the hive in an uncrowded area.

Starting with the mouthparts at rest, the mandibles are opened wide and the whole tongue is moved somewhat forward and downward. At the same time the distal portion of the proboscis is swung outward a little and a small droplet of nectar appears in the preoral cavity. The whole proboscis is then raised and retracted almost to the position of rest, but is depressed again, and is again raised as before, and so on. With each succeeding depression, the distal portion of the proboscis swings outward a little farther than before, but it makes only the beginning of a return to its position of rest.

Accompanying the second depression of the proboscis an increased amount of nectar appears in the preoral cavity, some of which begins to flow out over the upper surface of the tongue. As the proboscis is raised and retracted the second time, the beginning of a drop of nectar usually may be seen in the angle formed by its two major portions. This droplet increases in size each time the proboscis is alternately depressed and raised. The bee then draws the entire drop inside her body. As the nectar begins to be drawn in, the drop assumes a concave surface at its lower end.

Upon the completion of this part of the ripening process, the bee "searches" for a cell in which the drop she has been concentrating is deposited. This position is characteristic of a bee depositing unripe honey. If the cell is empty, she enters until her mandibles touch the upper rear angle of the cell. The nectar is forced out over the dorsal surface of the folded proboscis, between the mandibles, which are held well apart. Then, using the mouthparts as a brush, and turning her head from side to side, she "paints" the unripe honey across the upper wall of the cell so that it runs down and occupies the rear portion of the cell. But if the cell already contains honey, she dips her mandibles into it and adds her drop directly without the "painting" process.

When nectar is coming in rapidly, and particularly if it is very thin, the house bees do not always immediately put it through the ripening process, but deposit it almost at once.

Later these droplets are collected and are then put through the process of ripening by manipulation. Unripe honey is put through this phase of the ripening process repeatedly before it becomes fully ripened.

The honeybee changes the concentration of nectar very slightly while enroute to the hive.

Water is eliminated from fresh nectar and unripe honey.

NOTES

WHAT YOU NEED TO KNOW ABOUT *HONEY*

Among the many varieties of honey, several are outstanding — Tupelo honey is very much in demand because it does not granulate. When orange blossoms fill the air with sweet perfume, bees are busy making choice golden honey. Genuine orange blossom honey is light amber in color, heavy in body, and has the aroma of a grove in bloom. Gallberry honey is almost white in color, while Mangrove honey is very sweet.

In addition to the varieties of honey available to the consumer, there are also five types to choose from. These include liquid, comb, solid (sometimes called granulated or crystalized), chunk and cut comb.

Liquid honey is free of visible crystals. It is obtained by uncapping the combs and forcing the honey from the cells by centrifugal motion. It differs from comb honey only in the absence of the comb. In America, most people prefer honey in the liquid form.

Most honey is crystalized or granulated. Honey in this state, called "solid", is partially or wholly solidified or "sugared", and is often referred to as candied, creamed or spread. It is very popular in Canada and is finding an increasing demand in this country.

Comb honey, as its name indicates, is the honey in the comb as stored by the bees. Usually comb honey is served in its original size or cut into individual portions. This can be done by cutting the comb honey into one-inch squares with a knife dipped in boiling water. Chill thirty minutes before serving to prevent loss of honey from the comb. Serve the individual pieces with a cocktail.

More recently little chunks of sealed comb honey (about four inches long and one and one-half inches in width) have been wrapped in cellophane and packed in individual cartons. In the trade these are known as cut comb or honey hunks.

In the southern states, there is another type called chunk honey. The combs are built in shallow extracting frames and are cut in various sized chunks that will slip into tin pails and glass jars. The spaces between the combs and around them are filled with liquid honey.

NOTES

WHAT IS THE TRUTH ABOUT HONEY?

Honey is the only natural sweetener that doesn't have to be refined, and it is the only one that is not. Therefore, nothing is added or taken away, making it the world's only perfect sweetener.

There are as many colors and kinds of honey as there are plants from which the bee makes the honey, depending on the kind of flower from which the nectar was taken.

Honey differs chemically from any other sweetener. Honey contains minerals, nutrients, vitamins, acids and *natural* sugar, which table sugar or any artificial sweetener does not have.

Honey has no *"empty"* calories—only *"natural"* calories, everything the body needs to build and rebuild itself without adding weight.

Honey is nature's mystery. Bees do not make the honey from the nectar they gather by any chemical action. It is transformed into honey by enzymes produced in the honey sac, which converts the *natural* sucrose into honey. No one has identified all of the constituents of honey or knows exactly how it is made. Medical scientists have found that bacteria can't live in honey— while sugar breeds bacteria. Agricultural scientists have tested honey produced from plants heavily sprayed with pesticides and found that it never contains even a trace of any foreign chemical.

Nutritionalists point out that honey is the only food that includes all the substances necessary to sustain life, including water.

Laboratory analysis shows that honey is a supersaturated solution of glucose; 17% *natural* moisture and 79.5% various natural sugar, getting its sweetness from fructose. Most important for us, honey contains the remaining 3.5% proteins, minerals, vitamins, acids, "natural" calories, and natural sugars,—as well as the various flavor, aroma and pigment substances.

The amounts may seem small, but no other food contains all of these. The healthful properties of honey are outstanding for good health. A balanced amount of proteins, minerals, acids and natural calories are the main concern for a strong, healthy mind and body. These things were meant to be used in minute portions.

Honey contains enzymes which are used by the body to activate its chemical reactions; digestion being the number one reaction.

Honey will not cause fermentation because it is absorbed so quickly that distress cannot take place.

Honey is especially recommended for anyone whose digestion is bad.

Honey contains both glucose and fructose. Glucose is quickly absorbed by the body. Fructose is absorbed more slowly and is able to maintain blood sugar.

Glucose (also known as dextrose) is the body's main energy source and is the only sugar known to exist in its free state in the fasting human body.

"Sucrose" is refined white sugar (except where found in nature, in less than 5% total weight) the *empty* calorie type with the vitamins and minerals refined out.

Fructose (also known as levulose) has the same chemical formula as glucose but its structure is different. Fructose is found in free form in fruits.

Honey has the advantage over sugar as it doesn't create a higher blood sugar level than the body can assimilate.

Honey increases your energy because the honey is predigested by the honeybee in the bee's honey sac before you eat it. It goes straight to the bloodstream, turning to quick energy and goes through the bloodstream at a trickle.

Honey is excellent nourishment, and a power supply for the heart as well as the best food for quick energy.

Formula for infants made with honey retains both calcium and magnesium.

Honey is nature's only perfect, all-purpose sweetener. Man can neither better nor imitate honey in any way. Only the honeybee can make our oldest and most perfect sweetener.

Our natures are the physicians of our disease.

With the food value of honey, to any other food; no one can afford not to use honey.

If you leave the honey alone until the bees *fully* cap the honey, the honey will not ferment. The honey only ferments if you take it out before the bees finish taking the water out. They do this by fanning their wings until the water content is lowered to 16 to 17 percent water. Then they cap the honey. Until this is done, the honey is not finished or ripened.

Honey is more than just a sweetener, it is good nutrition.

Just studying honey and doing scientific studies does not make one a beekeeper or make one know all the facts or truths about honey. It takes more than just looking at it scientifically.

It takes years of working with the bees themselves, eating the honey, and seeing it for yourself.

It is not more likely that honey's digestible qualities are from its makeup, but it is a proven fact that honey is more digestible than any other sweetener.

Honey is made up of dextrose and levulose, *but* one advantage of using honey instead of sugar is that fructose (fruit sugar) makes honey sweet, *not* sucrose.

Honey has more levulose than dextrose so the body can handle it more easily, making it better for people over 55.

Honey increases your energy because the honey is predigested by the honeybee; it then goes straight to your bloodstream turning to quick energy. And honey doesn't turn to fat as sugars and artificial sweeteners do.

For the person who says that there is no difference between honey and sugar we would like to make it *very* clear that there is a *vast* difference between these two sweeteners, and to enlighten the misinformed. First, sugar which comes from beets or cane, is sucrose, a disaccharide, which overstimulates the islands or islets of Langerhans, which is very small cells located in the pancreas. The pancreas is a gland that secretes digestive juices and enzymes for metabolism of foods in the small intestines. Within this gland are the islands or islets of Langerhans; they are glands within a gland. The islands or islets of Langerhans furnish insulin used to burn up excess sugar.

When people eat *empty* calories, such as white sugar, white flour, white rice or sugared cereals, which do not contain nutrients, such as vitamins, minerals and enzymes, then the body must use the reserve B Complex vitamins, calcium and other nutrients to metabolize the empty calories. But if they continue to eat empty calories, they will run out of the reserve of these nutrients to metabolize the empty calories and then the islands or islets of Langerhans become overstimulated. If they continue to eat empty calories over a long period of time, the islands or islets of Langerhans become chronically overstimulated and they develop an ailment called hyperinsulinism. This means that the islands, which are now chronically overstimulated are secreting too much insulin and thus, they are burning up some of your own blood sugar. This is what causes low blood sugar or hypoglycemia. When an individual has low blood sugar, the brain is not getting all the energy it needs, because the brain lives solely on blood sugar—glucose. So any part of the nervous system can be affected.

Second, honey is not a disaccharide like sugar, it is a monosaccharide, which does not overstimulate the islands of Langerhans like sucrose does. Pure unadulterated honey can be tolerated in *small* amounts by diabetics and hypoglycemics. But sucrose causes trouble to a diabetic and hypoglycemic.

For the misinformed who still believes there is no difference between sugar and honey, *this is not so*. White sugar is an *empty* calorie, honey is not, but contains nutrients. In order to prove this—take a look at the nutrients that are in honey. In 3⅓ ounce serving or 100 grams of honey, we find 5 milligrams of calcium, 6 milligrams of phosphorus, 3 milligrams of magnesium, .5 milligrams of iron, 5 milligrams of sodium, .04 milligrams of B2, riboflavin, .3 milligrams of B3, niacin, and 1 milligram of Vitamin C. "You" be the judge.

51

NOTES

HINTS ABOUT HONEY

When To Use Honey—Use honey whenever you want to add a sweetly smooth and distinctive taste to your recipes. Honey has marvelous "keeping" qualities due to its ability to absorb and retain moisture which retards drying out and staling of baked goods. Cakes and cookies in which it is an ingredient stay fresh and moist much longer than those made without it. If you're baking goodies for children away at school or young men overseas in the armed forces or just friends half a continent away, honey will help keep them oven-fresh. Using it also means you can make party desserts well ahead of time and still be prepared to bow to the compliments of your guests.

How To Use Honey—In a cake or cookie recipe that calls for other sweetening, the general rule is to reduce the amount of liquid one-quarter cup for each cup of honey used. Honey may be substituted for sugar cup for cup. When honey is substituted in baked goods, add ½ teaspoon baking soda to the recipe for every cup of honey used and bake at a lower temperature.

Storing Honey—Honey should be stored in a tightly-covered container in a dark, cool place. Freezing or refrigeration will not harm the honey but may hasten granulation. If granules do form, place the jar of honey in a bowl of warm water (no warmer than the hand can stand) until all crystals are melted and honey is liquid. Too high a temperature will scorch the honey.

Types of Honey—Honey varies in color and flavor, depending upon the plant sources from which the bees gathered the nectar. It is generally classified and sold as liquid, comb, solid, chunk and cut comb.

LIQUID—is the most popular form because it has been forced from the comb and is free of crystals.

COMB HONEY—is sold in the comb as stored by the bees.

CUT COMB HONEY—is cut into small chunks and wrapped in individual cellophane covered cartons.

SOLID HONEY—is partially or wholly solidified or granulated and is referred to as candied, creamed or churned.

CHUNK HONEY is a combination of liquid and comb honey packaged in jars or pails.

Honey and Vitamins—Honey contains thiamine, riboflavin, ascorbic acid, pyridoxine, pantothenic acid, and nicotinic acid, all of which play vital roles in human nutrition.

Honey Contains Minerals—Among the mineral elements found in honey are iron, copper, sodium, potassium, magnesium, manganese, calcium and phosphorous. These elements are all essential to good nutrition.

NOTES

COOKING WITH HONEY

Honey is really the only perfect all-purpose sweetener. Honey doesn't overpower or intimidate the flavor of any food. Honey brings out the individual character of any meat. Honey will add a delightful touch to tea, coffee, fruit juices, cocktails, or to whatever you add it to. Honey will give a crisp distinct flavor to vegetables and salad ingredients. It also adds delicate touches to fruits and cereals. Pancakes, waffles, muffins or rolls are not complete without honey.

Wherever you use sugar or any artificial sweetener, you can and should use man's original sweetener—honey.

Foods cooked and sweetened with honey will have a better flavor and will keep your baked goods oven fresh for several days because of its marvelous keeping qualities.

One tablespoon of honey is equivalent to 5 tablespoons of sugar. Use ¾ cup of honey instead of one cup of sugar. Reduce total amount of other liquids by one quarter cup, per cup of honey.

When using honey in cooking, moisten the measuring spoon or cup first with water or oil, then measure the honey.

To neutralize honey's "natural" acidity, add ½ teaspoon of baking soda to the ingredients per cup of honey.

Store honey at room temperature, not in the refrigerator.

The real beauty of honey to all other sweeteners, is honey does more than just sweeten. Honey is nourishing, it adds its subtle depth, body and delicious flavor to any dish in which it is used, as well as a necessary ingredient that assures a moist and tender mixture in all breads, cakes and other baked goods that will help keep them moist for days.

All the recipes in this book have been prepared and taste-tested using honey. However, when substituting honey for sugar in your other recipes, follow these general guidelines:

★　★　★

Substitute ¾ cup of honey for sugar up to one cup. Reduce total amount of other liquids by one-quarter cup per cup of honey.

★　★　★

Lower baking temperature 40 to 50 degrees to prevent over-browning.

★　★　★

Foods sweetened with honey will have a better flavor if kept until the day after baking before it is served.

★　★　★

To bring crystallized honey back to its natural liquid state, place container of honey in a pan of warm water until crystals disappear. Set your electric dishwasher at 140 degrees.

NOTES

COOKING TIPS

No one food or group of foods, provide all the nutrients necessary.

Variety is a must in planning nutritious, meatless meals.

Raw foods in the vegetable and fruit groups yield more nutrients than processed.

Frozen vegetables without sauces and seasonings, and frozen fruits without added sugar, can be nutritious when fresh foods are too expensive.

Cooking is necessary to grain legumes, nuts, and seeds, to enable the body to digest the most nutrients from them.

Meatless diets emphasize whole foods, not refined foods.

Whole foods are recommended because there is a balance between the calorie and nutrient value that is lacking in refined foods. Nature packs the food supply with nutrients, but man has refined it into simple components without understanding how the digestive system creates a healthy nutritional balance in the body.

For balanced meatless meals try a choice of whole grain yeast bread, soybeans, kidney, garbanzo, or lima beans—sunflower seeds, almonds, walnuts for nutrients such as iron, thiamine, niacin, and protein.

For nutrients: protein, vitamin B12, riboflavin, calcium, phosphorus, and iron.

NOTES

TIPS

Honey and bee pollen use does not result in heavy production of body fat. It is palatable and digestible as well as nutritious.

★ ★ ★

Many nutrition experts consider bee pollen and honey excellent nourishment, a "power supply" for the heart muscle.

★ ★ ★

Besides being an unexcelled energy food, honey is one of nature's most powerful germ killers. Germs simply cannot survive in honey. In fact, primitive man not only used honey as food, but also as medicine to heal his wounds.

★ ★ ★

Mixed with lemon juice, honey is an excellent remedy for simple coughs.

★ ★ ★

Honey and bee pollen activate the taste buds and accentuate all the other flavors in any and all recipes.

★ ★ ★

For something mighty appealing to all the family, peel and slice oranges, dip the slices in honey and then in toasted shredded coconut. This is equally good served at breakfast time or offered as a salad at lunch, garnished with a few pecan halves.

★ ★ ★

Store honey at room temperature, not in the refrigerator. Keep container closed tightly and in a dry place.

★ ★ ★

If you're baking goodies for children away at school, young men overseas, or friends out of town, honey will help your baked goods stay oven fresh because of its marvelous "keeping" qualities.

★ ★ ★

For a quick "pick-me-up" try honey and one 130mm pod of bee pollen in milk. Since honey is 99 percent predigested, the energy from honey will move very quickly into the bloodstream, and will go through the bloodstream at a trinkle.

★ ★ ★

Honey will caramelize at a high temperature, so use a lower oven temperature (about 50 degrees lower) when baking with honey. In this way, a cake or bread cannot become too brown on top before it is done on the inside.

★ ★ ★

Use one-quarter teaspoon of baking soda for each cup of honey in baking. This will help neutralize honey's natural acidity.

★ ★ ★

Bee pollen and honey are natural laxatives, and two of the fastest working stimulants known.

★ ★ ★

Bee pollen and honey are natural unrefined foods, unique in that it is the only unmanufactured sweet available in commercial quantities.

Honey retains moisture to a greater extent than does sugar. Honey cakes and cookies will remain moist longer than those made with sugar.

★ ★ ★

Cover honey tightly because it tends to lose aroma and flavor and absorbs moisture when exposed to air.

★ ★ ★

Honey may darken slowly after many months, but it will still be the most nutritious food.

★ ★ ★

Keep honey in a warm, dry place where you would keep salt. Room temperature, 70 to 80 degrees is best.

★ ★ ★

If honey granulates, place container in hot water until it liquifies.

★ ★ ★

Nutrients found in bee pollen and honey are: Calcium, iron, phosphorous, sodium, thiamine, riboflavin, niacin and ascorbic acid.

★ ★ ★

Foods sweetened with honey will have a better flavor if kept until the day after baking before serving.

★ ★ ★

Freezing does not injure the color or flavor of honey, but may hasten granulation.

★ ★ ★

Honey spreads make the morning's toast extra special. Blend honey with softened butter, then add one of these: crushed strawberries, grated orange rind, or cinnamon.

★ ★ ★

Come cookout time, be sure you have an ample supply of honey on hand. It's dandy to use as a glaze on ribs, Canadian bacon or pork chops. Mix first with a bit of mustard. You'll like the added zip.

★ ★ ★

Add honey, mustard and a bit of ground ginger to canned baked beans and see what a difference it makes.

★ ★ ★

Honey over ice cream is delightful, but don't forget to drizzle a little over custard or raisin bread pudding, too.

★ ★ ★

Blend honey with peanut butter to create a fine glaze for baked ham. Great for sandwiches, too.

★ ★ ★

A honey sauce is scrumptious with baked chicken. One that's excellent is a mixture of one-quarter cup melted butter, one-quarter cup lemon juice. Use it for basting during the last half-hour of baking.

★ ★ ★

Whip canned sweet potatoes until light and fluffy. Add honey and cinnamon, spoon into a well buttered casserole and top with slices of pineapple to bake.

★ ★ ★

If you're out of salad dressing, toss crisp greens and fruit with honey. Tasty and easy!

★ ★ ★

Another dandy for fruit salads is easy and quick and very good. To make it, blend one-quarter cup honey with three-quarters cup dairy sour cream or yogurt.

★ ★ ★

A honey-mustard dressing can be used on either fruit salad or vegetables. While whipping one-half cup heavy cream until stiff, gradually add one tablespoon prepared mustard and one tablespoon honey. Chill before serving.

★ ★ ★

Two tablespoons of honey added to your favorite cake will make the cake wonderfully tender and less crumbly. For best results, add the honey in a fine stream to the batter as you beat.

★ ★ ★

A sweet-sour dressing is tart and refreshing on a vegetable salad. One-quarter cup honey, one-third cup vinegar and a dash of salt and pepper. Heat to boiling. Pour over salad and toss.

★ ★ ★

Add one tablespoon of honey and bee pollen to corn, carrots, green beans, etc. to bring out the true flavor to your vegetables.

★ ★ ★

For a quick and delicious cake icing try this: Warm solid honey slightly so that it will spread easily, add chopped nuts and coconut and gently spread over cake.

★ ★ ★

Honey-baked bananas are grand for dessert. Peel bananas, put them in a well-greased baking dish and brush with honey so that the entire banana is covered. Bake at 350 degrees for 15 to 20 minutes, or until the bananas are tender when tried with a fork.

★ ★ ★

Start the day off right with a good breakfast and be sure to include honey and bee pollen.

★ ★ ★

Try this one . . . when whipping cream, sweeten it with just a little honey. It will stay firm for much longer without separating.

NOTES

MEAD

Meads have been brewed by primitive tribes all over the world since the dawn of time. All are imbued with health-giving properties. And it is one of the curiosities of philology all over the world.

There are many varieties of mead renowned for their invigorating qualities.

Apparently there is a good deal of virtue in mead, for honey contains malic acid which counteracts gout and rheumatism, and this is transferred to the brew. Cider incidentally, also contains malic acid, and that is why it is often recommended specifically to rheumatic sufferers.

Mead is superior to grape wine, for it leaves no hangover; is a blood-cleanser and acts as an invigorating tonic.

For many centuries mead was considered a veritable elixir vitae. Its principal medicinal value was in kidney ailments as an excellent diuretic without disastrous effects upon the kidneys.

TRUE MEAD

For whatever amounts of mead desired, boil 3 parts water to one part honey over a slow fire until full one-third of the "must" has evaporated. Carefully skim the mixture and store in a cask for 3 to 4 days before serving.

HONEY MEAD

Use 3 lbs. of mild-flavored, fine quality, extracted honey, 2 gallons of water, 2 lemon peels for each gallon of water and boil the mixture 30 minutes. The lemon peel is added as the mixture boils. Skim well. Work this mixture with yeast and place in a cask for five or six months. Then bottle for use. If you choose to store the mead for several years, add 4 lbs. of honey to each gallon of water.

Although different recipes vary widely in their requirements, the following might be kept in mind as general brewing suggestions:

1. Use soft water—clean, pure rain water or distilled water for brewing.
2. Use vitreous enameled or tinned utensils, or stainless steel, or copper vessels. Earthenware or glass utensils may be used but honey "must not", and should never, at any stage of the brewing, come into contact with iron, brass or galvanized metal containers.
3. Use only the very best mild-flavored honey to produce fine vintage mead.
4. The yeast largely determines the character of the mead. Use pure-culture wine yeasts such as Maury, Madeira, or Malaga. Brewer's yeast will give the mead a peculiar tang.

5. The "must", or basic solution of honey and water, must be sterilized and any possible contamination avoided. Also sterile vessels or casks in which the mead is fermented should first be sterilized.
6. Keep the temperature constant (65-70 degrees F.) during the fermentation process.
7. The best time to begin fermentation is during the summer months (May-June) as temperatures and the atmospheric conditions are most favorable to the preparation and growth of yeast cells.
8. Mature the mead in oaken casks. It takes great patience to produce fine wines and mead will be of excellent quality if stored in oaken casks seven years before bottling, although it can be used earlier.

While the flavor, bouquet, and character of the mead are determined by the floral sources of the honey used and the type of yeast, the body and alcoholic content of the mead is largely determined by the proportion of honey to water used. A mead of less than 2 lb. of honey to a gallon of water will not keep while the strongest concentration that can be effectively fermented is 6 lb. of honey to gallon of water.

THE NO DIET WAY

You can take off twelve pounds a month by this *No Diet* method—no gimmicks or fad diets involved.

Balanced nutrition is the only way to take off weight and keep it off without hurting your health—you can lose up to twelve pounds a month until you reach your desired weight—then stay there, by eating and enjoying life.

It's not how much you eat—but what you eat, and when you eat.

Not by counting calories but by the kind of calories you eat.

The perfect weight and health balanced way to eat and live is especially for *you*—not a rat or a monkey in a lab.

It has been tested by many people who now have no weight problems.

Start off the day by looking at yourself (nude) in a full length mirror.

For Breakfast
(your most important meal)

1 glass of fresh orange juice

1 cup of natural grain cereal

1 slice of whole wheat toast with polyunsaturated margarine or unadulterated honey (if you crave sweets)

1 cup of coffee (if you must have a sweetener use 1 teaspoon of unadulterated honey)—no cream or artificial creamer.

Change from orange juice to any other fresh juices (no concentrates)

Change your cereal as desired.

Switch your breakfast several times a week.

½ grapefruit—a peach—pear or fresh fruit as desired.

1 egg, cooked any way except fried

1 small bowl of old-fashioned oatmeal or cracked wheat (if you must have a sweetener, 1 teaspoon of unadulterated honey)

Use skim milk.

Lunch

1 cup of vegetable or fish soup

1 tomato or carrot—small amount of lettuce with vinegar and lemon juice

1 open face tuna or salmon sandwich on whole wheat bread. About 3-oz. of the fish mixed with 1 tablespoon of mayonnaise, thinned with fresh lemon juice and as much celery as you like.

1 glass of skim milk—1 glass of water—try and skip coffee

Snacks — "whenever" you get hungry

Fresh apple, fresh orange, grain wafers, celery, carrot, banana, grapes, skim milk banana milk shake, orange juice, tomato juice (fresh), popcorn (unsalted), cup of apple juice, *hot or cold*, stone ground wheat crackers.

Dinner

3-oz. roasted or broiled (skinless) chicken or your favorite fish

½ cup of brown rice with chopped parsley—or broccoli or carrots or beets or corn on the cob or ½ cup green beans

6-oz. of broiled lean steak (with all fat cut off)—twice a week

1 or 2 glasses of water

1 cup of apple cider—Hot or Cold

NOTES

BEESWAX

The bee makes the wax for its comb. Wax is a by-product of metabolism. To make one pound of wax the bee has to consume six or seven pounds of honey.

Each cell is perfectly formed and consists of six sides, the angles of which are always the same, the only exceptions being cells fashioned to fill in odd corners. The hexagonal cell is made so accurately that man could adopt the cell of the hive as a standard measure. Everything the bee does is mathematically correct and logical.

Bees make their cells in such a way that they tip up, ever so slightly, so that the honey will not drain out by gravity; nor will their cells melt until the temperature reaches 140 degrees F.

NOTES

GLUCOSE

Blood sugar is glucose, the fuel your body runs on. And when glucose runs low, every part of your body can start to sputter and slow down. Muscles ache, vision blurs, and digestion can clog. These are physical problems.

The mind is powered by the brain, and the brain, too, is part of the body. And the brain, more than the muscles, or more than any other organ demands glucose. It has to have glucose, and it has to have it now. Low blood sugar can cause any mental problem. Nervousness, anxiety, irritability, depression, forgetfulness, confusion, indecisiveness, poor concentration, and nightmares.

Faced with this flood of glucose, the pancreas pumps out extra insulin—so much insulin, that even some of the glucose that should circulate in the body is swept out of the bloodstream. The result: low blood sugar—and low spirits.

When blood sugar drops, the brain sends out an SOS. The adrenal glands come to the rescue pumping harmones into the blood that release the glucose stored in the liver and muscles.

"This sudden burst of blood sugar triggers the pancreas to release too much insulin, and blood sugar drops."

"Low blood sugar is epidemic in America."

It's never too late—or too early to deal with low blood sugar.

NOTES

FRUCTOSE

What is fructose?

Fructose is the sweetest of all sugars, found in fruits of all kinds, berries, vegetables and honey—It is the main natural sweetener in honey. Honey has natural calories—not empty calories (the fatting kind).

"WHAT IS NATURAL FRUCTOSE?"

Honey contains *"Natural Fructose"* 39% of natural fruit sugar which is also known as levulose. Honey—contrary to the action of glucose which *restores* oxygen and is replaced by lactic acid when fatigue sets in, *natural fructose, found in honey*—crystallizes more readily and is used for tissue rebuilding.

Natural fructose is the sweetest of all sweeteners and is found in fruits of all kinds, berries, vegetables and *Honey natural fructose is not white* in color.

Honey has all the *natural* fructose—*natural* vitamins, minerals, amino acids, natural vitamins—everything the body needs including water—without the need of one artificial coloring or *man made* food.

FRUCTOSE

Do not make the *mistake* that the fructose you buy in the stores is the same as you find in *Honey—fruit* or *juice*.

The name is now *misleading* says Dr. Theron Randolph. "Fructose", was formerly referred to as "fruit sugar",—what you are buying in the stores is *now made* from *corn starch* . . . (and) allergy to corn is the leading *cause of allergy* in this country, when you try it and find that rather than relieving your fatigue, *this fructose increases* it. *"Many* of the harmful effects of sucrose are due to this *fructose components"* (the average 140 grams intake of sucrose yield about 70 grams of this fructose). According to a report from the Journal of the American Medical Association, "This" fructose may be converted to excessive amounts of sorbital which the Russians consider to be a possible factor in the eye hemorrhages which are so frequent in severe diabetes.

Dr. Palnis studies, have found that this fructose raises the level of triglyceride in the blood. If you use this fructose, make sure to test your triglyceride level.

This new *fructose*—can be bought now in tablets or in powdered form by jar or box at about $3.00 a pound. Looking like white refined sugar, it is less sweet.

According to Carlton Fredericks, from a report published in Britain's *Lance*, this fructose causes increased uric acid formation and inhibition of protein synthesis. This raises blood lactate and serum acid to undesirable. In disorders involving the liver or lack of oxygen in the tissues, infusion of *this fructose* leads to lactic acidosis.

According to Dr. Milton M. Heshel, Chief of Endocrinology and Metabolism at Einstein Medical Center Northern Division, this "fructose" is readily converted to glucose by the liver and does raise blood sugar levels.

Diabetic specialist Dr. Philip Gerber agrees, saying, "I would not recommend it for any diabetics."

HONEY OR SUGAR

For the person who says there is no difference between honey and sugar, we would like to make it **very** clear that there is a **vast** difference between these two sweeteners, and to enlighten the misinformed. First of all, sugar, which comes from beets or cane, is sucrose, a disaccharide, which over-stimulates the islands or islets of Langerhans, (masses of very small cells) which are located in the pancreas. The Pancreas is a gland that secretes digestive juices and enzymes or islets of Langerhans, glands within a gland. These islands or islets of Langerhans furnish insulin to burn up excess sugar.

When people eat **empty** calories, such as white sugar, white flour, white rice or sugared cereals, which do not contain nutrients, such as vitamins, minerals and enzymes, then the body must use the reserve B complex, calcium and other nutrients to metabolize the empty calories. But if they continue to eat empty calories, they will run out of the reserve of these nutrients to metabolize the empty calories and then the islands or islets of Langerhans become overstimulated. If they continue to eat empty calories over a long period of time, then the islands or islets of Langerhans become chronically overstimulated and they develop an ailment called hyperinsulinism. This means that the islands, which are now chronically overstimulated are secreting too much insulin and thus burning up some of your own blood sugar. This is what causes low blood sugar or hypoglycemia. When an individual has low blood sugar, the brain is being kept from all the energy it needs, because the brain lives solely on blood sugar - glucose. So any part of the nervous system can be affected.

Second, honey is not a disaccharide like sugar; it is a monosaccharide, which does not overstimulate the island of Langerhans like sucrose does. Pure unadulterated honey can be tolerated in small amounts by diabetics and hypoglycemics.

What is meant by unadulterated honey? So that there can be no misunderstanding, pure unadulterated honey is levulose or fructose, which is a monosaccharide. Bees do not manufacture sucrose.

Beekeepers remove honey in the spring and in the fall. Whenever you remove honey, you must leave enough honey for the bees to live on. Bees make more honey than they have to have so when you take honey from the hive, it is called "robbing the hive."

However, some beekeepers take too much of the honey from the beehives, and the bees starve in the winter unless the beekeeper feeds them syrup or sugar water.

Honey was made by the bees for the bees and a good beekeeper will always leave enough honey for the bees to live on.

For the misinformed who still believe there is no difference between sugar and honey, **this is not so.** White sugar is an **empty** calorie; honey is not, but contains nutrients. In order to prove this, take a look at the nutrients that are in honey.

In 3⅓ ounces or 100 grams of honey, we find 5 milligrams of calcium, 6 milligrams of phosphorus, 3 milligrams of magnesium, 5 milligrams of iron, 5 milligrams of sodium, .04 milligrams of B2, Riboflavin, .3 milligrams of B3, Niacin, and 1 milligram of Vitamin C. YOU be the judge.

NOTES

WHY DO BEES INVERT SUCROSE INTO NECTAR?

Honey is made by bees by the process of the collecting of fairly thin concentrations of sugar solutions, called nectar (many are sucrose), "adding an enzyme" which converts the sucrose to glucose and fructose, and then evaporating the mixture to a concentration of 80% solids for storage. Plants are glucose/fructose disacharride sucrose for energy transport, this is why a little of the sugars glucose, fructose and sucrose are available to honeybees, to be the main source and only one controlled for pollination progress.

Why do bees invert the sucrose into nectar? Because honeybees need to maximize stability of the stored syrup, which must be high enough to prevent bacteria and yeast growth but not so high that crystallization will occur. Sucrose solutions will crystallize readily at concentrations which are osmotically sterile. The most stable mixture is 50% glucose and 25% fructose (dry basis) honey, being 100% invert.

HMF

When honey is overheated, fructose is partially turned into hydroxymethylfurfural (HMF). According to all standards, honey should not be put on sale unless its HMF content is lower than 4 mg/100g. HMF content exceeds 10mg/100g **only** when honey is adulterated. Honey which has not been heated is the only honey which has no HMF.

NOTES

BEES IN WINTER

Temperature exerts an important influence upon the activities of the honeybee. Normal honeybee activities usually occur at temperatures between 10 degrees and 38 degrees. At temperatures closely approaching 38 degrees or above, bees seldom forage, except for water, and remain within the hive or cluster on the outside. A single inactive bee soon loses the ability to fly at 10 degrees Centigrade and at temperatures below 7 degrees C. soon becomes motionless. But the honeybee colony possesses the ability to maintain and regulate its own temperature to a remarkable degree. In an active brood nest every bee acts as a thermostat. When the temperature in the brood area falls below approximately 35 degrees C. the heating process starts in the thoraces of bees, increasing the temperature to the normal level. In the broodless winter cluster the temperature of the thoraces of bees ranges between 20 degrees and 36 degrees C., independently of the outside temperature. Normally it stays around 29 degrees C.

Bees wander at irregular intervals in the cluster, eat honey, and participate in heating activity. A single bee having enough food can maintain considerably higher temperatures than her surroundings. Bees are able to produce heat during long periods without any muscular movements.

When the temperature immediately surrounding the bees falls to 14 degrees C. or lower, they form a cluster. The winter cluster when first formed is usually located in the lower part of the hive, often near the front. Throughout the course of the winter it moves upward and to the rear of the hive. By spring the cluster in a 2-story hive is found most often in the upper story. It has been noticed, however, that when the cluster was formed in the fall a part of it, consisting of 100 to 200 bees, always protruded over the end bars of the frames. This "connective cluster" shifted with the movement of the main cluster. When the bees began to move toward the rear of the hive, the "connective cluster" on the end bars of the frames disappeared and simultaneously a new one was formed on the top bars of the frames.

At relatively warm temperatures the bees on the surface of the "connective cluster" are almost motionless except they sometimes quickly move their wings. Their wings are somewhat extended so that they overlap each other, forming a kind of cover as if preventing the loss of heat. At low temperatures the bees perform frequent and quick movements with their wings, as if shaking them; the lower the temperature the more often they move their wings. During the severe cold the bees on the surface of the cluster bury their heads and thoraces inside the cluster so that only the abdomens are visible, and at the same time they move their abdomens with screwlike motions. When at such times one listens at the entrance, using a special hearing device resembling a physician's stethoscope, one can hear screaky noises suggestive of a procession of carts with unoiled axles. As soon as it becomes warmer the bees spread out again. The temperature inside the hive may be quite low.

The "connective cluster" plays an important part in the life of bees in winter. Quite often one can see single bees moving from one interspace to another. There are openings toward the inside of the cluster and frequently, especially during the morning, a bee or two can be seen ventilating on the edges of such openings. The moisture from the air of the cluster condenses on the wall of the hive and sometimes is collected and utilized by the bees. During the winter the bees "scrub" the bars with "rocking movements." Sometimes they clean each other. At this time they perform a "grooming" dance.

There are losses of bees during the winter; bees may die from various causes - those that left the cluster because of need of defecation; bees which somehow remained behind on the combs when the cluster contracted; or those which left the cluster for some other cause. These losses in some cases may be extremely severe.

GATHERING AND STORING WATER

Water is gathered and is used primarily for diluting honey as larval food, and for cooling and humidifying the hive interior. Water is required by the nurse bees whenever it becomes necessary to thin honey in the processing of larval food. When fresh nectar is available, it usually is used without dilution in the preparation of brood food. Hence, the activity of water carriers is most noticeable in early spring before the honeyflow begins, and that it practically ceases when nectar is coming into the hive in abundance, unless the nectar is highly concentrated. The extent to which water is needed by adult bees has not been determined. Workers or queens caged on candy readily take water when it is offered to them and live longer than do those that lack water. This would indicate that adult bees also need water.

When a water carrier brings her load of water into the hive and climbs on the comb, she begins vigorous dancing. Usually four or five bees follow each dancer, and at more or less frequent intervals the dancer pauses long enough to transfer a sip of water to one of the nearby workers. At times a water carrier dances for a full minute before transferring her load. Sometimes a water carrier enters, "performs" a brief dance, then proceeds rapidly to dispose of her load. Sometimes she gives a small amount to each of a half-dozen bees in quick succession before resuming her dance, and then, after dancing awhile, transfers the balance of her load to one or two bees. It is not unusual to see two or three bees being supplied all at one time by a single water carrier.

Having unloaded, she begins preparation for her next field trip by securing a small amount of food from one or more of the house bees or she goes to a cell and takes honey. Then she almost invariably gives her tongue a swipe between her front feet, rubs her eyes, often cleans her antennae, and then leaves quickly. These preparations and the quick exit are so characteristic that an observer soon is able to tell whether a bee is leaving for the field, or whether she is just going to another part of the hive.

NOTES

THE HONEYBEES BIOLOGICAL CLOCK

THE DANCE COMMUNICATION has been extensively studied and is widely accepted. Inside a hive, directional data supplied during the tail wagging portion of dances is dependent on the sun for a source of orientation. Obviously, the sun is constantly moving, subsequently requiring a constant revision of the directional data supplied by the dancer. Herein lies one of several uses of the honeybee's biological clock. Bees change the position of the sun in their calculations after as little as 7-11 minutes. In other research, results were obtained that indicated this time may be as short as five minutes, but no less than four.

Not only are bees able to constantly revise their data, but they are able to make the revision without having recently seen the sun. This observation was derived from marathon dancers. These dancers are actually scout bees that have been searching for a homesite before a swarm has been cast. These bees are capable of dancing for hours (to include night hours) without having seen the sun, all the while maintaining accurate sun location.

NOTES

THE QUEEN

The queen bee is the central figure of the colony. All life revolves around her. Without a queen, there can be no community of bees. Only she has the power to reproduce the species, so she is a slave to her destiny, which is a practical one of producing an endless supply of eggs. In her productive span of approximately four years, she is capable of producing a million new bees. This ceaseless production becomes a constant challenge to her workers, for it is up to them to provide the so called brood cones. Nature has long since taught the bees how to create a new queen, and now we are about to witness a marvel without parallel in the wild, one of nature's most astounding secrets of life.

After a brief search among the ordinary brood cells, the worker bees select a half a dozen of the most recent larve. These are the candidates for queen, and for each a royal cradle is constructed. The queen is the largest of the bees (and an aspirant put out on her) must have room to grow into her position. And now the magic potion of this fairy tale. Like sorcerers with supernatural power, the nursemaids produce a substance called royal jelly, and when this is fed to the regal babies, it works a miracle, for it makes them fertile. Upon reaching maturity, any one of them will be able to recreate the species. Finally the queen cells are sealed up. A period of waiting begins. The fate of the colony depends on the outcome of this mysterious ritual. At long last, the queen appears.

About five days after emergence the virgin makes convulsive movements with her abdomen and at the same time the abdominal orifice opens for several seconds, signifying that the virgin is sexually mature. In the midday hours on the day of mating the number of excited workers around the queen increases. Some of them run toward the entrance and one can often see a continuous line of bees between the virgin and the entrance. Finally a group of bees fanning, with their scent glands exposed, gather in front of the entrance and normal flying and foraging almost ceases.

The virgin appears at the entrance accompanied, and often appears to be driven there, by several workers. If for some reason the queen is hesitant to fly, workers will prevent the queen's return into the hive and "try to force her" to flight. So the mating of the queen not only involves the queen and the drones, but the whole colony participates in the event.

The drone alights on the back of the abdomen of the queen, grasping her with all six legs, his head extending over the thorax. The abdomen of the drone curls downward until it contacts the tip of the queen's abdomen. As soon as ejaculation occurs, the instantly paralyzed drone releases from the queen and topples over backwards, and 2 to 3 seconds later a distinct "pop" is heard as the two separate. If this explosion fails to occur, they may fall to the ground where they separate within a few minutes. Drones are strong fliers and are capable of carrying the queen along with them in flight.

The duration of nonmating flights ranges from 2 to 30 minutes, and that of mating flights from 5 to 30 minutes.

SWARMING

Swarming is the means of reproduction of colonies. By swarming, the number of colonies is increased, replacing those that perish from adverse living conditions.

Early in the season the bees are in a relatively compact cluster in the hive. As the season advances, the bees begin to occupy more and more space. At any given temperature there are a certain number of bees per unit area present on the brood combs. Because the young bees have a tendency to remain in the warmest place, the newly emerged bees remain on the brood combs and the older bees move out. Older bees do not move far, but remain on the combs adjacent to the brood area where they begin to clean cells.

When the queen comes to such a comb, she starts to lay immediately and the young displaced bees begin to feed her and the larvae that subsequently appear. Therefore, the egg-laying activity of the queen is governed by the displaced nurse bees since she can lay only as many eggs as there are cells prepared by the nurse bees.

In a normal colony the queens's retinue consists of 10 to 12 nurse bees surrounding the queen in a more or less closed circle, leaving some space between them and the queen. They constantly touch the queen, especially her abdomen, sometimes licking it. During intensive laying, the "rest" periods of the queen last 10 to 15 minutes, at which time she receives food from several bees.

During the swarming season, before the queen cells are started, there is an increased activity in egg laying. In one case the queen deposited 62 eggs in 45 minutes, i.e., 1,968 eggs per day. The circle around the queen is formed in close proximity to the queen and the bees in the circle become excited. They constantly and persistently offer food to the queen, even pushing their heads under the head and thorax of the queen.

Several days before swarming, an abnormal number of bees may be seen "resting" quietly on the bottom of the combs. At this time, the searchers may begin to look for a new nesting place. The searchers perform the wagtail dance in the hive indicating the direction and the distance to the prospective new site. The nest searchers, contrary to the food gatherers, do not interrupt the dance but continue dancing for hours, or even days, changing the direction of their dance according to the change in the position of the sun.

The number of bees going out with the swarm may be from about half to 90 percent of the parent colony.

When a swarm settles it consists of two distinct parts: an outside shell about three bees thick, quite compact, and the inner, loose part, consisting of chains which are connected with the shell in many places. The shell gives protection to the swarm against outside influences and provides the swarm with necessary mechanical strength. In the shell there is a distinct entrance to the inside of the cluster.

There is a division of labor in the swarm cluster. The searchers are all over 21 days old; the bees in the shell are 18 to 21 days of age; and those inside the cluster are house bees, up to the age of 18 days. The bees in the shell are constantly changing places; during a 10-minute period two-thirds of the surface bees changed places with those of the inside portion of the shell.

BEESWAX

As the main building material, wax is always in demand. How the bees manufacture wax is one of the honeybee's most astounding secrets. During the wax making, they hang suspended in a living curtain. This is called the time of waiting while a strange chemistry produces the wax inside their bodies. As soon as the wax appears in the form of thin flakes, squeezed from between the abdominal segments, the other workers gather up the flakes and chew them to a workable consistency and begin the construction of the honeycomb.

Little pearly discs of wax, somewhat resembling fish scales protruding from between the rings on the underside of the body of the bee, reveal little wax scales of rare beauty. These scales come so fast that they fall on the bottom board of the hive and may be scraped up in large quantities. During the season where a colony has plenty of room, wax scales are seldom wasted.

The bees remove these wax scales from their bodies and construct the comb. The wax scales are scraped off by one of the large joints of the hind leg, the spines piercing and catching into the scale; then the leg, by a peculiar maneuvering, is moved up to where the forelegs grasp the scale. The scale is manipulated in the mandibles, where it is applied to the comb. During this process, the bee stands on three legs while the other hind leg and the two forelegs perform the wonderful sleight of hand by which they transfer the scale from one portion of the body to the other.

The delicate architecture of the bee is an amazing example of engineering to conserve space. Each cell with its precision-made six walls fits all other cells surrounding it. This hexagonal structure with built-in cross bracing assures maximum strength with a minimum of material. Every hive contains hundreds of cones and during the spring and summer they're packed with golden honeybee pollen from the flowers and nature's most perfect foods — nature's golden treasure, Honey.

ROYAL JELLY

Royal Jelly is secreted by hypopharyngeal glands of worker bees normally 5-15 days of age. It is fed to queens throughout their larval and adult lives, and also to young workers and drone larvae. Being high in protein, it is synthesized during the digestion of pollen, although it is believed that honey is also added to the secretion. Royal Jelly is described as a creamy, milky white, strongly acidic, highly nitrogenous substance, with a slightly pungent odor and a bitter taste.

NOTES

NOTES

PROPOLIS

Propolis is as old as bees themselves, and honeybees have been making and using it since the beginning of time.

The first recorded use of propolis was by the Greeks who named the substance by combining the words "PRO" (before) and "POLIS" (city). The Greeks suspected that the bees used propolis to defend their "city" against bacterial and other harmful intruders.

What is propolis?

Bees use propolis for their own health. Without some form of antibiotic protection, 60,000 bees crawling over one another in a densely crowded hive would die of bacterial infections.

Bees discovered that certain trees exuded a bud sap used by the trees themselves to fight infection. Through the centuries the bees have taken this sap back to the hive, added their own secretions to it, and distributed the metabolized "antibiotic" in every corner and comb in the hive. Some scientists believe that the inside of a hive is more sterile than a modern hospital.

The power of propolis to fight infection in the hive and keep the bees free from bacterial disease is amazing. For instance, if any object gets into the hive and is stung to death by the bees, it would normally decay, causing serious bacterial problems. However, the bees cover the object with propolis and "mummify" the object. It remains embalmed without decay.

Hippocrates (460-377 B.C.), the father of medicine, used propolis to help heal ulcers and externally as a salve, to treat wounds and sores.

The famous Roman naturalist, Pliny (23-79 A.D.) wrote in his Historia Naturalis, that propolis "disposed tumors, moderated nerve illnesses, and healed ulcers."

Nicholas Culpepper's famous complete herbal states that propolis as an ointment "is singularly good for all inflammations in many parts of the body and cools the heat of wound . . ." One of the most recognized uses of propolis in modern history was an ointment to treat wounds during the Boar War.

Now the medical world is faced with another dilemma — how to react to bee propolis. This natural food supplement, with its antibiotic properties, has astonished doctors.

Just what are the raw qualities of propolis? What does it contain? It contains resin, vitamins, amino acids, minerals and is rich in a natural selection of bioflavonoids. It also possesses a whole natural selection of micro-elements in trace amounts.

Propolis has proven effective in helping to deal with a wide variety of infections and illnesses, including ulcers, colds, stress, hypertension, pharyngitis and periodontal problems.

Propolis helps to maintain the body's general health.

Dr. John Diamond, president of the International Academy of Preventive Medicine, in his book Behavioral Kinesiology says, "Our life energy is the source of our physical and mental well-being, of glowing health, of the joy of living . . ." Unfortunately 95 percent of the general population tests low on the life energy scale. He points out that the activity of the thymus gland helps control one's life energy. "I have never," he continues, "seen a patient with a chronic degenerative illness who did not have an underactive thymus gland. I believe it is the thymus weakness, or underactivity, that is the original cause of the illness.

". . . Of all the natural supplements I have tested the one that seems to be the most strengthening to the thymus, and hence the life energy, is bee resin, or bee propolis — a resin secreted by trees and then metabolized by the bees . . ."

By activating the thymus gland, propolis can help to keep a body in good health. It helps to build up the body's resistance to disease.

When propolis is taken internally, the rate of metabolism is increased and the resistance of the organism to the influence of unfavorable factors from the outside world is raised . . . It follows that the mechanism for healing in propolis is based not only on its antibacterial propensity and the detoxifying effect, but also on the influence of defensive reactions of the organism, and propolis - in contrast with antibiotics - intensifies the whole immunological reactive capability of the micro-organism.

Essentially, propolis acts to stimulate the immune system of the body.

Dr. Remy Chauvin, the world's leading authority on bee therapy says, "Today with the over-use of antibiotics, bacteria is developing a marked resistance to their purported benefits. Infections are on the increase and some diseases - formerly not contagious - are becoming highly transmittable.

"Scientists believe that nature has a cure for every disease. It's just a matter of finding it. With the introduction of propolis, it is possible that we can one day abolish most drug-related chemicals. Also remember that keeping the body free from disease through natural healing can actually slow down the aging process and add years to the life span."

The antibacterial and antiviral properties of propolis work to raise the body's natural resistance to disease by internally stimulating one's own immune system. In doing this, propolis also supplies added amounts of vitamins B, C, and K, and all essential minerals including iron, calcium, aluminum, manganese, and silicon.

Other normal antibiotics do the opposite - they can kill the friendly bacteria in the system which the body uses to synthesize vitamins B and K. This can cause a severe deficiency of these important vitamins.

Propolis is the natural daily supplement food which is the ultimate preventive "medicine."

The International Beekeeping Technology and Economy Institute says, "Propolis has antiproteolytic bactericidal and bacteriostatic action that is unparalleled among natural substances.

The uniqueness of propolis is dramatically described by an Australian scientist, E. L. Ghisalberti of the Department of Organic Chemistry at the University of Western Australia.

"It is not the normal drug, it is a natural supplement . . ."

A propolis gatherer, after finding the source of propolis, bites immediately into it with her mandibles and tries, with the assistance of the first pair of legs, to tear a little piece off. She may knead the latter a little between her mandibles and then, with the help of one of the second legs, she quickly transfers the piece of propolis to the pollen basket on the same side. She may do this in a stationary position or in flight. Next, the bee places a piece of propolis in the pollen basket on the other side. The clumps of propolis are patted frequently with the second leg, putting them in proper shape. Such alternate loading continues until both baskets are fully loaded. To make a load, the bee is active from 15-60 minutes.

When a bee enters the hive with a load of propolis, she is unloaded by another worker bee that bites at the propolis and pulls, tears off a little piece, then presses it firmly into place. While applying propolis, the "cementing bee" may mix some wax into it. The propolis carrier immediately makes the load smooth again by patting. She may be freed of her load in the course of an hour, or several hours, depending on the use of propolis in the hive. When bees have been freed of their load, they immediately forage for more. Propolis is collected on warm days only. Propolis gatherers remain "faithful" to their work, but there are very few of them in each colony. Bees of foraging age collect propolis.

Unlike existing antibiotics, propolis is not produced by micro-organisms. Its antibiotic activity is dependent to a larger extent on flavonoids that originate from higher organisms, i.e., their evolutionary age is about the same as humans.

Specifically, the flavonoids in propolis come from plants which have a major part of their metabolism in common with human beings and are, therefore, in most respects mutually compatible.

One of the world's most respected experts in the area of flavonoids and propolis is Professor Bent Hausteen, formerly of Cornell University and now at the School of Medicine, Kiel University in West Germany.

Propolis is a resinous substance. It is greenish-brown to chestnut in color. It has a very pleasant smell of poplar buds and honey - beeswax and vanilla.

It contains 55 percent resins and balms - 30 percent wax - 10 percent etheric oils and 5 percent pollen. Still many chemical compositions of propolis have never been identified.

Propolis is rich in biologically active vitamins, especially the B group, minerals and for anti-bacterial activity flavonoids.

Short-lived antibiotic drugs can scarcely compare with propolis which has been used by nature-believing beekeepers since God created bees when He created all things.

Propolis toothpaste is my favorite toothpaste.

NOTES

BEE VENOM

The components of bee venom have been reviewed and summarized in several papers. Bee venom is quite complicated chemically. It contains several biochemically or pharmacologically active substances, including at least the following: histamine, dopamine, melittin, apamin, mast cell destroying (MCD) -peptide, minimine, and the enzymes phospholipase A, and hyaluronidase. There are at least eight protein fractions of honeybee venom, of which phospholipase A, melittin, and apamin are the three major ones. Bee venoms from different areas of the country at various times of the year contain the same complement of proteins, which would indicate that bees synthesize these independent of pollen source.

Bee venom is a clear liquid with a sharp, bitter taste, an aromatic odor, an acid reaction, and a specific gravity of 1.1313. It dries quickly at room temperature, to 30-40 percent of the original liquid weight.

Honeybee venom is more toxic than wasp venoms. In rare instances, one sting is capable of causing death by anaphylactic shock to persons who are hypersensitive. Such persons may die within 30 minutes, unless prompt medical relief is given, which normally consists of administering ice packs, adrenalin, and antihistamines. Under more normal circumstances, however, it would probably require at least 500 stings over a short period of time to cause death by direct toxicity. One person in Africa is said to have received 2,000 stings and survived.

BEE STING

When the honeybee stings it injects protein and various other chemical substances. When the sting area swells, it's because of an allergic reaction to the foreign protein. It may swell and itch like a gnat bite for a short time and then disappear. This is the normal reaction of about 90 percent of people who have never been stung before.

Beekeepers gradually acquire a resistance to stings in a very short time.

If you are allergic to a bee sting, go to an emergency room or doctor immediately.

USES AND POTENTIAL USES

Bee venom may have potential use in medical practice. Some interest has been expressed in introducing a pure bee venom product on the American drug market, to be used in two ways: (1) for the treatment of rheumatoid arthritis, and (2) for the desensitization of hypersensitive individuals.

The idea of using bee venom in the treatment of rheumatoid arthritis is centuries old, and is based at least in part on the observation that beekeepers are seldom afflicted. "Bee venom therapy," or "apitherapy" has long been practiced in Europe, and in recent times has been advocated in the United States. To date however, the effectiveness of the use of bee venom for this purpose has never been definitely established; the problem has never been properly investigated in an unbiased manner.

The effectiveness of bee venom in the desensitization of hypersensitive individuals is currently under investigation. The interested reader should refer to the paper by Barr (1971) for a review of the world literature on the subject. Although the injection of whole insect extracts are currently being used for the purpose of desensitization (in part because pure venom has not been available in quantity), they contain many proteins that are chemically unrelated to those found in pure venoms.

"You Are Someone Special And Should Treat Yourself As Such

Taking care of yourself and your eating habits should not be treated as though you were a rabbit, a rat, a monkey or guinea pig in a test tube laboratory. You should know facts, not just test experiments or go on "theory".

In these times of crazy, faddish diets, teenagers, grown-ups, older people, and people of all walks of life may be shocked to discover it has been and is still perfectly normal for the people, who since the beginning of time have been eating bee pollen and honey. Using propolis, beeswax, and bee venom for preventive medicine.

True natural food is not old-fashioned or old wives' tales, but has been proven down through the ages to be more effective than any synthetic food or drug of so called modern food.

NUTRITION

The modern science of nutrition discovered that body heat is the result of oxidation within the tissues. More than forty substances are essential for human nutrition. The first essential for good nutrition is enough food to supply energy for muscular work and to maintain normal weight. Honeybee pollen will give you more balanced nutrition in minute form.

An inadequate intake of energy food results in underweight, which leads to lack of endurance and lowers resistance to disease. Excessive amount of fat in the body overtax the heart, interfere with the function of other essential organs, and reduce life expectancy.

Protein is the material of which all cells are made. There are many varieties of proteins, each made up of a mixture of simpler units, amino acids, and nutritional value, depending on the assortment of amino acids.

Minerals are required by the body, such as calcium, phosphorous, iron, iodine — other trace minerals are needed in **minute amounts.**

Vitamins are substances which are vital to life. More than a dozen vitamins have been identified and at least six are known to man to be essential for human health and welfare.

Honeybee pollen and honey are the only foods known to have all of these in minute amounts that you need for a balanced everyday diet. These things were meant to be in minute quantities.

Proper nourishment is necessary for good health. Our body is made of living cells. Cells are born and they have a natural life span. Every cell is a complex chemical factory.

FATS, PROTEINS, CARBOHYDRATES

The three main ingredients of any diet are fats, protein, and carbohydrates. To stay alive you need a certain amount of calories, which you can get from any one of these three ingredients. You can only get so many of your calories in proteins before you begin to develop serious problems from excess protein intake. To eat a very low fat diet, you wind up needing a very high amount of carbohydrates in your diet.

All carbohydrates are made up of combinations of sugar molecules. When you eat carbohydrates, no matter what kind they are, your body soon breaks them down into sugar molecules.

There is a world of difference between how the body handles large complexes of sugar molecules (that is, complex carbohydrates) and its handling of sugar all by itself. A carbohydrate molecule is nothing other than a group of sugar molecules. Some carbohydrate molecules are very small, consisting of only a few sugar molecules. Sucrose, for example, is a carbohydrate molecule consisting of only two sugar molecules, a glucose molecule (blood sugar) tied to a fructose (fruit sugar). Sucrose is what we ordinarily refer to as "table sugar".

Carbohydrates molecules consisting of only one, two, or a very few molecules are called simple carbohydrates. Table sugar, molasses, and syrup are all examples of simple carbohydrates. All the other carbohydrates are called complex carbohydrates and typically are made up of very, very large molecules. Honeybee pollen again will help you to balance your protein, fats and carbohydrates.

To keep them going, the cells need energy. The cell can get energy from any one of three of the types of chemicals; from fats, from amino acids or from carbohydrates.

Carbohydrates are a preferred energy source. It is important that the body have fats and amino acids. When food is plentiful and there are more carbohydrates available than the body needs to use for energy, it will convert some of the carbohydrates to fat and store them. These fat reserves can be called into play to give the energy needed to keep the cell factories going. If fat reserves are used up, our bodies can then begin to use body protein as a last resort, breaking it down into amino acids and getting the necessary energy from the amino acids themselves.

The body definitely needs natural sucrose, glucose and fructose but at complex starch rates, not at candy bar rates.

"DISEASE HAS ONLY NATURAL CAUSES"

Our body contains millions of cells. Each cell is a tiny power plant, that uses food as fuel and changes it to energy.

Energy gives you the power to think, to move, to read, to write, to laugh, and to cry.

The heart must have energy to pump blood, the chest muscles, to cause breathing. Every move you make requires energy, even blinking your eyes.

Some parts of the body use energy at all times. Twenty four hours a day — even when you sleep, you are using energy.

The only way the body can get all this energy is by the food we eat. Food provides the building materials the body must have to grow and to repair itself.

Proteins furnish building blocks for the body. The body does not store protein. You should have proteins everyday so the body does not get soft and flabby.

Carbohydrates are stored by the body in the form of starch in the muscles changing to fat. The fat is stored all over the body under the skin.

Water is not a food but you cannot live without it. Water forms the basis of your blood and of your tissue fluids — keeps the body temperature even — a must for your daily diet.

Nature intended that we have a trickle of glucose sugar passing through out intestinal wall at all times, and one teaspoon of glucose sugar in our bloodstream at all times. This is ultra-critical to our well being. Honeybee pollen everyday — at least 5 mm will help you have the energy you need 24 hours a day.

TOO RUN DOWN WHEN NIGHT COMES?

In order to feel amorous and full of energy you must eat and drink the right foods.

A good breakfast with food packed with energy is the sure way to start the day.

Your whole body runs on "energy," all day and all night. When you get tired and nervous, eat. It's a sign the body needs energy. Snacks are good for you, as long as they are not junk; fresh orange juice, apples, other fruits — as long as the food value has not been taken out or refined, until there is no energy value left. Make sure the food you eat at snack time or at meal time isn't loaded with chemicals or artificial sweeteners. Pollen for example, is loaded with all the things the body needs for energy and well being. Try keeping a jar of honey at work, in your car or where you can see it at home, while you work.

Food loaded with such things as nitrates or another name for potassium, saltpeter; we all know what that will do for you.

Pollen is the most valuable ally in any program to partake fully of the ex-lixir of human life.

Nature intended that we have a trickle of glucose passing through our intestinal wall at all times and one teaspoon of glucose in our bloodstream at all times.

This is ultra-critical to our well being.

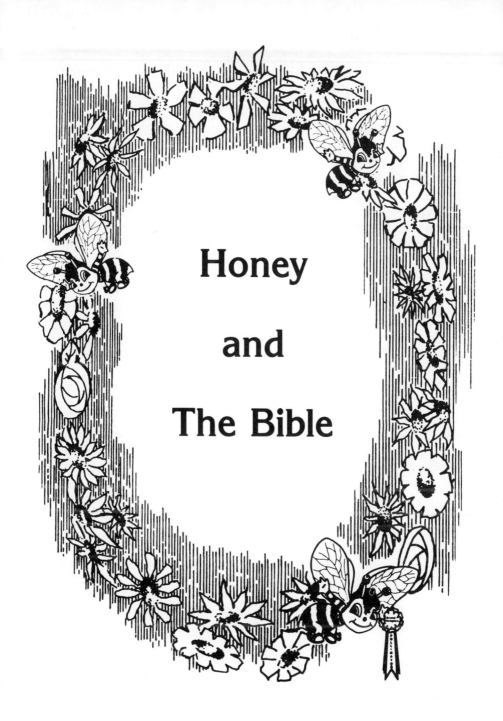

Honey

and

The Bible

Genesis 43:11 — And their father Israel said unto them, If it must be so now, do this; take of the best fruits in the land in your vessels, and carry down the man a present, a little balm, and a little honey, spices, and myrrh, nuts, and almonds:

Exodus 3:8 — And I am come down to deliver them out of the hand of the Egyptians, and to bring them up out of that land unto a good land and a large, unto a land flowing with milk and honey; unto the place of the Canaanites, and the Hittites, and the Amorites, and the Perizzites, and the Hivites, and the Jebusites.

Exodus 3:17 — And I have said, I will bring you up out of the affliction of Egypt unto the land of the Canaanites, and the Hittites, and the Amorites, and the Perizzites, and the Hivites, and the Jebusites, unto a land flowing with milk and honey.

Exodus 13:5 — And it shall be when the Lord shall bring thee into the land of the Canaanites, and the Hittites, and the Amorites, and the Hivites, and the Jebusites, which he sware unto thy fathers to give thee, a land flowing with milk and honey, that thou shalt keep this service in this month.

Exodus 16:31 — And the house of Israel called the name thereof Manna: and it was like coriander seed, white; and the taste of it was like wafers made with honey.

Exodus 33:3 — Unto a land flowing with milk and honey: for I will not go up in the midst of thee; for thou art a stiffnecked people: lest I consume thee in the way.

Leviticus 2:11 — No meat offering, which ye shall bring unto the Lord, shall be made with leaven: for ye shall burn no leaven, nor any honey, in any offering of the Lord made by fire.

Leviticus 20:24 — But I have said unto you, Ye shall inherit their land, and I will give it unto you to possess it, a land that floweth with milk and honey: I am the Lord your God, which have separated you from other people.

Numbers 13:27 — And they told him, and said, We came unto the land whither thou sentest us, and surely it floweth with milk and honey; and this is the fruit of it.

Numbers 14:8 — If the Lord delight in us, then he will bring us into this land, and give it us; a land which floweth with milk and honey.

Numbers 16:13 — It is a small thing that thou hast brought us up out of a land that floweth with milk and honey, to kill us in the wilderness, except thou make thyself altogether a prince over us?

Numbers 16:14 — Moreover thou hast not brought us into a land that floweth with milk and honey, or given us inheritance of fields and vineyards: wilt thou put out the eyes of these men? we will not come up.

Deuteronomy 6:3 — Hear therefore, O Israel, and observe to do it; that it may be well with thee, and that ye may increase mightily, as the Lord God of thy fathers hath promised thee, in the land that floweth with milk and honey.

Deuteronomy 8:8 — A land of wheat, and barley, and vines, and fig trees, and pomegranates; a land of oil olive, and honey.

Deuteronomy 11:9 — And that ye may prolong your days in the land, which the Lord sware unto your fathers to give unto them and to their seed, a land that floweth with milk and honey.

Deuteronomy 26:9 — And he hath brought us into this place, and hath given us this land, even a land that floweth with milk and honey.

Deuteronomy 27:3 — And thou shalt write upon them all the words of this law, when thou art passed over, that thou mayest go in unto the land which the Lord thy God giveth thee, a land that floweth with milk and honey; as the Lord God of thy fathers hath promised thee.

Deuteronomy 31:20 — For when I shall have brought them into the land which I sware unto their fathers, that floweth with milk and honey; and they shall have eaten and filled themselves, and waxen fat; then will they turn unto other gods, and serve them, and provoke me, and break my covenant.

Deuteronomy 32:13 — He made him ride on the high places of the earth, that he might eat the increase of the fields; and he made him to suck honey out of the rock, and oil out of the flinty rock.

Joshua 5:6 — For the children of Israel walked forty years in the wilderness, till all the people that were men of war, which came out of Egypt, were consumed, because they obeyed not the voice of the Lord: unto whom the Lord sware that he would not shew them the land, which the Lord sware unto their fathers that he would give us, a land that floweth with milk and honey.

Judges 14:8 — And after a time he returned to take her, and he turned aside to see the carcase of the lion: and, behold, there was a swarm of bees and honey in the carcase of the lion.

Judges 14:9 — And he took thereof in his hands, and went on eating, and came to his father and mother, and he gave them, and they did eat: but he told not them that he had taken the honey out of the carcase of the lion.

Judges 14:18 — And the men of the city said unto him on the seventh day before the sun went down, What is sweeter than honey? and what is stronger than a lion? And he said unto them, If ye had not plowed with my heifer, ye had not found out my riddle.

I Samuel 14:25 — And all they of the land came to a wood; and there was honey upon the ground.

I Samuel 14:26 — And when the people were come into the wood, behold, the honey dropped; but no man put his hand to his mouth: for the people feared the oath.

I Samuel 14:27 — But Jonathan heard not when his father charged the people with the oath: wherefore he put forth the end of the rod that was in his hand, and dipped it in an honeycomb, and put his hand to his mouth; and his eyes were enlightened.

I Samuel 14:29 — Then said Jonathan, My father hath troubled the land: see, I pray you, how mine eyes have been enlightened, because I tasted a little of this honey.

I Samuel 14:43 — Then Saul said to Jonathan, Tell me what thou hast done. And Jonathan told him, and said, I did but taste a little honey with the end of the rod that was in mine hand, and, lo, I must die.

II Samuel 17:29 — And honey, and butter, and sheep, and cheese of kine, for David, and for the people that were with him, to eat: for they said, The people is hungry, and weary, and thirsty, in the wilderness.

I Kings 14:3 — And take with thee ten loaves, and cracknels, and a cruse of honey, and go to him: he shall tell thee what shall become of the child.

II Kings 18:32 — Until I come and take you away to a land like your own land, a land of corn and wine, a land of bread and vineyards, a land of oil olive and of honey, that ye may live, and not die: and hearken not unto Hezekiah, when he persuadeth you, saying, The Lord will deliver us.

II Chronicles 31:5 — And as soon as the commandment came abroad, the children of Israel brought in abundance the firstfruits of corn, wine, and oil, and honey, and of all the increase of the field; and the tithe of all things brought they in abundantly.

Job 20:17 — He shall not see the rivers, the floods, the brooks of honey and butter.

Psalm 19:10 — More to be desired are they than gold, yea, than much fine gold: sweeter also than honey and the honeycomb.

Psalm 81:16 — He should have fed them also with the finest of the wheat: and with honey out of the rock should I have satisfied thee.

Psalm 119:103 — How sweet are thy words unto my taste! yea, sweeter than honey to my mouth!

Proverbs 24:13 — My son, eat thou honey, because it is good; and the honeycomb, which is sweet to thy taste.

Proverbs 25:16 — Hast thou found honey? eat so much as is sufficient for thee, lest thou be filled therewith, and vomit it.

Proverbs 25:27 — It is not good to eat much honey: so for men to search their own glory is not glory.

Isaiah 7:14, 15 — Therefore the Lord himself shall give you a sign; Behold, a virgin shall conceive, and bear a son, and shall call his name Immanuel. Butter and honey shall he eat, that he may know to refuse the evil, and choose the good.

Isaiah 7:22 — And it shall come to pass, for the abundance of milk that they shall give he shall eat butter: for butter and honey shall every one eat that is left in the land.

Jeremiah 41:8 — But ten men were found among them that said unto Ishmael, Slay us not: for we have treasures in the field, of wheat, and of barley, and of oil, and of honey. So he forbare, and slew them not among their brethren.

Ezekiel 3:3 — And he said unto me, Son of man, cause thy belly to eat, and fill thy bowels with this roll that I give thee. Then did I eat it; and it was in my mouth as honey for sweetness.

Ezekiel 16:13 — Thus wast thou decked with gold and silver; and thy raiment was of fine linen, and silk, and broidered work; thou didst eat fine flour, and honey, and oil: and thou wast exceeding beautiful, and thou didst prosper into a kingdom.

Ezekiel 16:19 — My meat also which I gave thee, fine flour, and oil, and honey, wherewith I fed thee, thou hast even set it before them for a sweet savour: and thus it was, saith the Lord God.

Ezekiel 20:6 — In the day that I lifted up mine hand unto them, to bring them forth of the land of Egypt into a land that I had espied for them, flowing with milk and honey, which is the glory of all lands.

Ezekiel 20:15 — Yet also I lifted up my hand unto them in the wilderness, that I would not bring them into the land which I had given them, flowing with milk and honey, which is the glory of all lands.

Ezekiel 27:17 — Judah, and the land of Israel, they were thy merchants: they traded in thy market wheat of Minnith, and Pannag, and honey, and oil, and balm.

St. Matthew 3:4 — And the same John had his raiment of camel's hair, and a leathern girdle about his loins; and his meat was locusts and wild honey.

St. Mark 1:6 — And John was clothed with camel's hair, and with a girdle of skin about his loins; and he did eat locusts and wild honey.

Revelation 10:9 — And I went unto the angel, and said unto him, Give me the little book. And he said unto me, Take it, and eat it up; and it shall make thy belly bitter, but it shall be in thy mouth sweet as honey.

Revelation 10:10 — And I took the little book out of the angel's hand, and ate it up; and it was in my mouth sweet as honey: and as soon as I had eaten it, my belly was bitter.

Solomon 4:11 — Thy lips, O my spouse, drop as the honeycomb: honey and milk are under thy tongue ; and the smell of thy garments is like the smell of Lebanon.

Solomon 5:1 — I am come into my garden, my sister, my spouse: I have gathered my myrrh with my spice; I have eaten my honeycomb with my honey; I have drunk my wine with my milk: eat, O friends; drink, yea, drink abundantly, O beloved.

Luke 24:40-43 — And when he had thus spoken, he shewed them his hands and his feet. And while they yet believed not for joy, and wondered, he said unto them, Have ye here any meat? And they gave him a piece of a broiled fish, and of an honeycomb. And he took it, and did eat before them.

Whether therefore ye eat, or drink, or whatsoever ye do, do all to the glory of God. (I Corinthians 10:31.) If we eat incorrectly it will harm us physically, and this is wrong.

God has given us our bodies, and they are to be used for His glory. If we neglect or abuse our bodies, we will not be able to serve him as we should. We should be concerned about our health. Our bodies actually belong to God if we give our lives to Christ.

Don't you know that you are God's temple and that God's Spirit lives in you? If anyone destroys God's temple, God will destroy him. (I Corinthians 3:16-17.)

Do you not know that your body is the temple of the Holy Ghost which is in you, which ye have of God, and ye are not your own? For ye are bought with a price: therefore glorify God in your body, and in your spirit, which are God's. (I Corinthians 6:19-20.)

If we are spiritually fit, we must be concerned about the body God has given us and wants us to use to his glory.

And also that every man should eat and drink, and enjoy the good of all his labour, it is the gift of God. I know that, whatsoever God doeth, it shall be for ever: nothing can be put to it, nor anything taken from it: and God doeth it, that men should fear before him. (Ecclesiastes 3:14-15.)

For every creature of God is good, and nothing to be refused, if it be received with thanksgiving: for it is sanctified by the word of God and prayer. If thou put the brethren in remembrance of these things thou shall be a good minister of Jesus Christ, nourished up in the words of faith and good doctrine whereunto thou has attained.

BEEKEEPING

There is no other activity that produces as much fun and pleasure as a hobby, or as much food for the amount of money.

A beehive (colony of bees) cost approximately $100.00. With proper care and handling, you can make $100.00 worth of honey the first year.

There are about six million hives in the United States. Recreational beekeeping is increasing as never before. The popularity of honey is also increasing. If you want to get started in beekeeping you will find a beekeepers organization in every state, most counties, all of Canada and probably every country in the world.

For beekeepers across the country that you can depend on to furnish you with pure honey — following is a few from coast to coast. There are many, many others, but there's just not room enough to mention them all. These are the larger ones and may help you realize how large the honey industry really is.

Adee Honey Farms, P.O. Box 368, Bruce, SD 57220 — 605-627-5621

Ahrens Bee Supply, 2279 Central Rd., Everson, WA 98247 — 206-966-5568

A. I. Root Co., 1700 Commerce Rd., Athens, GA 30607 — 404-548-7008

A. I. Root Co., P.O. Box 706, Medina, OH 44258 — 216-725-6677

American Bee Supply Inc., P.O. Box 555, Lebanon, TN 37087 — 615-444-7908

Beaver Valley Apiaries, 805 - 3rd Ave. West, Owen Sound, Ont. N4K 4P4
   John or Janet Toews

Betterbee Inc., Box 37, Rte. 29, Greenwich, NY 12834 — 800-833-4343

Cook-Dupage Beekeeper's Assn. Inc., 1 S. 235 Pine Lane, Lombard, IL 60148 —
   312-627-0443

Seldon Cooper, Rt. 1, Weaversville, NC 28787 — 704-645-4782

Cox Honey Farms, 456 S. State, Shelley, ID 83274 — 208-357-7874

Dadant & Sons Inc., Hamilton, IL 62341 — 217-847-3324
   Charles Dadant

D & D Honey, 1864½ Davis St., Elmira, NY 14901 — 607-733-5238

Dakin Farms, Rt. 7, Ferrisburg, VT 05456 — 802-877-2936

Desert Honey Co., 3113 E. Columbia St., Tucson, AZ 85714 — 602-746-1084

Drapers Super Bee, Rt. 1, Millerton, PA 16936 — 800-233-4273

Drapers Super Bee, Rt. 3, Auburn, NE 68305 — 402-274-3725

El Paso Bee Club, 10220 Bermuda Ave., El Paso, TX 79925 — 915-592-3355

Glorybee Honey & Supplies, 1001½ Terry St., Eugene, OR 97402 — 503-485-1649

Golden Crown Apiaries, 4001 Springfield Rd., Vandalia, OH 45377 — 513-898-5201

Grayson's Honey, Rt. 1, Box 1113, Owassa, OK 74056 — 918-272-3465

Haefelis Honey Farms, 0041 S. Rd. 1 E, Monte Vista, CO 81144

Happy Hive, 4476 Tulane, Dearborn Hgts, MI 48125 — 313-562-3707

Hives & Honey, SR Box 80632, Fairbanks, AL 99701 — 901-488-6484

Hob Bee World,17448 So. Center Park Way, Tukwila, WA 98188

Honey Bee, 6753 Hwy. 290 W., Austin, TX 78749 — 512-892-3515

Honey Center, 116 Neal St., Grass Valley, CA 95945 — 916-273-6424

Honey Heaven, 128 E. 11th Ave., Eugene, OR 97401 — 503-344-5939

Hubbard Apiaries, M-50 at Springville, P.O. Box 160, Onsted, MI 49265 — 517-467-2051

Darrell Jester, P.O. Box 4, Osceola, AR 72370 — 501-563-5701

F. W. Jones & Son, Ltd., Bedford, Quebec, Canada J0J 1A0 — 514-248-3323

Walter T. Kelly Co., Clarkson, KY 42726 — 502-242-2012

Keystone Honey Farms, 4375 Blacklick Rd., Pataskalo, OK 43062 — 614-837-0916

Lane Honey Bee Farm, 2172 Old Garden Valley Rd., Roseburg, OR 97470 — 503-672-2763

Lewis Co. Beekeepers, 340 Forest Napa Vine Rd. E, Chehali, WA 98532 — 206-262-3721

C. L. Maserang Apiaries, 4036 Kearby St., Ft. Worth, TX 76111 — 817-834-0721

W. F. Mathes Bee Farms, Rt. 1, Box 120, Pleasant Hill, MO 64080 — 816-987-2566

Miller Honey Co., 125 Laurel, Colton, CA 92324 — 714-825-1722

Mitchell Brothers Honey, 728 N. Davis St., Missoula, MT 59801 — 406-549-7644

N. C. State Beekeepers, 307 N. Holden Rd., Greensboro, NC 27410 — 919-299-6540

Pen. State Beekeeping Assn., Rt. 2, Box 310, Dumas, PA 18222 — Mary Young

Pierce Co. Beekeepers Assn., Rt. 2, Ellsworth, WI 54011, Evelyn Christens, Treas.

Queen Bee Gardens, 1863 Lane 11½, Lovell, WY 82431 — 307-848-2543

Red Deer Bee Supplies, 3904 - 51 St. Close, Red Deer, Alberta — 343-1093

Ruhl Bee Supply, 1424 N.E. 80th Ave., Portland, OR 97213 — 503-256-4231

Thelma Saxby, 7742 Riverview, Kansas City, KS 66112 — 913-299-8072

Gary Schmidt, Rt. 2, Box 6, Martin, SD 57551 — 605-685-6528

W. Stoller's Honey Inc., Box 97, Latty, OH 45855 — 419-399-3232

Strauser Bee Supply, 3rd & C Street Airport, Box 991, Walla Walla, WA 99362 —
    509-529-6284

Sunstream Bee Supply Center, P.O. Box 225, Eighty-Four, PA 15330 — 412-222-3330

Tropical Blossom Honey Co., P.O. Box 8, 106 N. Ridgewood, Edgewater, FL 32032 —
    904-428-9027

Warmuth Honey House, 17262 Sierra Hwy., Canyon Country, CA 91351 — 805-252-2350

Weaver Apiaries Inc., Rt. 1, Box 256, Navasota, TX 77868 — 713-825-2312

Jo Ann Weber, Rt. 2, Clayton, WI 54004 — 715-948-2732

York Bee Company, Inc., Harvey York, Jr., Jesup, Ga.

If you or your organization need a program, movie slides or lecture program on the job of cooking or canning with honey, write or phone:

The Wonderful World of Honey
Honey, Inc.
Rt. 2
Berryville, AR 72616
Phone (501) 423-3131

NOTE: Refer to Appendix A for a list of some beekeepers across the country that you can depend upon for pure honey.

INDEX

INDEX

The Wonderful World of Honey

A SUGARLESS COOKBOOK

*How Honey Helps You in Every Day Living
*Facts You Should Know About Nutrition
*The Miracle Healer
*Honey as a Beauty Food
*Honey as a Preservative

by Joe Parkhill

THE WONDERFUL WORLD OF HONEY . . .

. . . Teaches **You** how to change any recipe to a healthy, delicious, "sugarless" recipe. The popularity of honey is increasing as never before, here is the newest and fastest selling Sugarless Cookbook covering the subject of honey.

It contains over 300 taste-tested recipes; featuring salads, vegetables, meats, breads, cookies, candies and desserts. All "natural" good and delicious with honey — plus HEALTH HINTS, BEAUTY IDEAS and FACTS YOU SHOULD KNOW about honey and the history of honey.

It took over 25 years to gather these priceless recipes from beekeepers and nutrition-minded people from all over the world. These recipes are not just for a few desserts — or as some honey recipes, call for sugar and honey, these recipes are for everyday living and **this book** contains recipes for every day's entire menu.